Paper Chemistry

To Lesley Joy and Jennifer Anne

PAPER CHEMISTRY

Second edition

Edited by

J.C. ROBERTS

Professor of Paper Science
University of Manchester Institute of Science and Technology

BLACKIE ACADEMIC & PROFESSIONAL

An Imprint of Chapman & Hall

London · Glasgow · Weinheim · New York · Tokyo · Melbourne · Madras

Published by
Blackie Academic & Professional, an imprint of Chapman & Hall
Wester Cleddens Road, Bishopbriggs, Glasgow G64 2NZ

Chapman & Hall, 2–6 Boundary Row, London SE1 8HN, UK

Blackie Academic & Professional, Wester Cleddens Road, Bishopbriggs, Glasgow G64 2NZ, UK

Chapman & Hall GmbH, Pappelallee 3, 69469 Weinheim, Germany

Chapman & Hall USA, 115 Fifth Avenue, Fourth Floor, New York NY 10003, USA

Chapman & Hall Japan, ITP-Japan, Kyowa Building, 3F, 2-2-1 Hirakawacho, Chiyoda-ku, Tokyo 102, Japan

DA Book (Aust.) Pty Ltd, 648 Whitehorse Road, Mitcham 3132, Victoria, Australia

Chapman & Hall India, R. Seshadri, 32 Second Main Road, CIT East, Madras 600 035, India

First edition 1991
Second edition 1996

© 1996 Chapman & Hall

Typeset in 10/12 pt Times by AFS Image Setters Ltd, Glasgow
Printed in Great Britain by St Edmundsbury Press, Bury St Edmunds, Suffolk

ISBN   0 7514 0236 2

A catalogue record for this book is available from the British Library
Library of Congress Catalog Card Number: 95–80783

∞ Printed on acid-free text paper, manufactured in accordance with ANSI/NISO Z39.48-1992 (Permanence of Paper).

Preface to the second edition

Although the title of this book is *Paper Chemistry*, it should be considered as a text about the chemistry of the formation of paper from aqueous suspensions of fibre and other additives, rather than as a book about the chemistry of the raw material itself. It is the subject of what papermakers call wet-end chemistry. There are many other excellent texts on the chemistry of cellulose and apart from one chapter on the accessibility of cellulose, the subject is not addressed here. Neither does the book deal with the chemistry of pulp preparation (from wood, from other plant sources or from recycled fibres), for there are also many excellent texts on this subject. The first edition of this book was a great success and soon became established as one of the Bibles of the industry. Its achievement then was to collect the considerable advances in understanding which had been made in the chemistry of papermaking in previous years, and provide, for the first time, a sound physico chemical basis of the subject. This new edition has been thoroughly updated with much new material added.

The formation of paper is a continuous filtration process in which cellulosic fibres are formed into a network which is then pressed and dried. The important chemistry involved in this process is firstly the retention of colloidal material during filtration and secondly the modification of fibre and sheet properties so as to widen the scope for the use of paper and board products.

As is the fashion these days, each chapter is written by an internationally recognised expert in the field, and my thanks are extended to all of the contributors for their hours of patient and unseen research during the preparation of their manuscripts.

The introductory chapters on the surface chemistry and electrokinetics of the papermaking system are designed to lay the foundation for the understanding of later chapters in which surface chemistry and colloidal phenomena are of central importance. Following on from these, I wanted to address the fundamental aspects of polyelectrolytes particularly with respect to their use as retention aids, and have included an additional chapter at this point. The subsequent chapters deal with retention aid chemistry and our current understanding of the mechanism of action of polyelectrolytes and with paper modification by chemical addition. The principles of the chemical systems available for wet- and dry-strength improvement, the chemistry of control of water penetration (sizing) in both acidic and basic conditions, dyeing and optical brightening, and pigment particle addition for the modification of optical properties are discussed. The measurement and control of the chemical processes is the subject of the penultimate chapter, and the final chapter deals with practical applications of paper chemistry under real conditions.

The book is designed to accentuate the chemical principles involved in the papermaking system rather than to be a compendium of additives and, as such, it should be useful to the paper-mill chemist and to students and researchers in the field. Those chemical industries supplying the paper industry should find the book a useful reference source.

J.C.R.

Acknowledgement

The editor wishes to acknowledge the contributions made to the original edition of this book by Ms E.W. Hughes (editorial assistance), and Mrs E. Rodgers and Mr Z. Yajun (for assistance with some of the diagrams).

Contributors

Dr C.O. Au — Paper Chemicals Division, Eka Nobel Ltd. Blackburn, Lancashire BB1 5RP, UK

Dr R. Bown — English China Clays Research Centre, St. Austell, Cornwall PL25 4DJ, UK

Dr N. Dunlop-Jones — International Business Sector – Paper, Clariant UK Ltd, *formerly* Sandoz Chemicals (UK) Ltd, Horsforth, Leeds LS18 4RP, UK

Dr J.M. Gess — Weyerhauser Company, Tacoma, Washington 98477, USA

Dr D. Horn — Department of Polymer Physics, Polymer Research Division, BASF, 67056 Ludwigshafen, Germany

F. Lafuma — Laboratoire de Physico-Chimie Macromoléculaire de l'Université Paris VI — CNRS URA 278 — ESPCI, 10 rue Vauquelin 75231, Paris Cedex 05, France

Dr T. Lindström — Mo Och Domsjö AB, Research and Development, S-891 80 Örnsköldsvik, Sweden

Dr F. Linhart — Applications Department Paper, Marketing Colorants and Process Chemicals, BASF, 67056 Ludwigshafen, Germany

Dr J. Marton — Consultant and Adjunct Professor, 8 Northrup Court, Newtown, Pennsylvania 18940-1854, USA

Dr S.G. Murray — International Business Sector – Paper, Clariant UK Ltd, *formerly* Sandoz Chemicals (UK) Ltd, Horsforth, Leeds LS18 4RP, UK

Professor F. Onabe — Head of the Laboratory of Pulp and Paper Science, University of Tokyo, Bunkyo-ku, Tokyo 113, Japan

Dr G.A.F. Roberts — Department of Fashion and Textiles, The Nottingham Trent University, Nottingham, UK

Dr J.C. Roberts — Department of Paper Science, UMIST, PO Box 88, Manchester M60 1QD, UK

Contents

7 Wet-strength chemistry

N. DUNLOP-JONES

8 The sizing of paper with rosin and alum at acid pHs

J.M. GESS

9 Neutral and alkaline sizing

J.C. ROBERTS

10 Dyes and fluorescent whitening agents for paper   161
S.G. MURRAY

11 Physical and chemical aspects of the use of fillers in paper   194
R. BOWN

1 Applications of paper chemistry

J.C. ROBERTS

1.1 Introduction

Pulp and paper manufacture spans a very wide range of science and technology. It embraces plant genetics, anatomy, physiology, chemistry, biochemistry, physics, mathematics and computation and it involves considerable chemical, mechanical, electrical and process engineering. It is also an integral part of the subject of material science. The chemistry of the process might therefore seem, on the face of it, to be a relatively small and unimportant part of the overall technology. Nothing, however, could be further from the truth. Correct control of the chemistry of all aspects of the process is vital if product quality and process efficiency are to be maintained. Incorrect operation can lead to serious quality deficiencies or production losses and, finally, to lost markets.

The chemistry of pulp and paper manufacture can be considered to fall into four distinct areas, namely pulping, recycling, paper formation and finishing. This book is concerned with the chemistry of paper formation from aqueous suspensions of fibres, fillers, and other furnish components. More familiarly, it is the subject of wet-end chemistry.

Papermakers use many different synthetic and natural chemical additives for a variety of different reasons during the wet formation process. They may be used to influence the efficiency of the formation process itself or to improve final sheet properties. They are usually added (with the exception of fillers) at a level of around 0–5% by weight of the furnish and, because of their relatively high cost in comparison to fibre and filler, they often represent a significant proportion of the total raw material costs. The cost-efficient optimisation of their use can represent a source of considerable savings to a paper manufacturing company.

The most important area of application as far as the formation process is concerned is in the retention of fines and the dewatering and consolidation of the wet web. Additives used for this function are often polymeric and are frequently charged. They rely for their effect, as do many additives, upon adsorption to particle surfaces and on the influence that such adsorbed molecules have upon the state of flocculation of the dispersion. The physical chemistry and electrokinetics of these adsorption processes are complex and not well understood, although a great deal of progress has been made in this area over the past twenty years. The subject has been well reviewed recently [1] and is discussed in detail in Chapters 3 and 5.

Additives which are used for product modification may vary very widely in molecular type. They may be inorganic pigments which, because of their small particle size and high light scattering power, are used as opacifying agents, or they may be relatively simple small organic molecules as in the case of many sizes and dyestuffs. They may also be complex polymers as, for example, dry-strength additives, and they may or may not be soluble in water. In the latter case they require to be prepared as an aqueous suspension (e.g. as a sol or emulsion). However, if they are to influence paper properties, they must be efficiently retained in the wet web, and the chemistry of their retention is of key importance.

Chemicals which are added at the wet end for the purpose of the modification of product properties are usually intended to influence the bulk properties of the sheet, whereas those added at the size press or during coating are intended to influence predominantly surface properties. In many cases it is necessary to modify the bulk properties of the sheet even though retention of the chemical component in the wet web may be difficult.

For the period from around 1840 to the early 1970s paper was usually made in an acidic environment at pHs of around 4–5. This was because many grades required the use of rosin and aluminium sulphate for sizing, that is to render the sheet resistant to the penetration of aqueous fluids (see Chapter 8). Solutions of aluminium sulphate, because of the dissociation of the hexahydrated aluminium ion in aqueous solution, exhibit a pH of around 4.5. Aluminium sulphate has also been popular with papermakers because it behaves as a mildly effective retention aid. The reasons for this are explained more fully in Chapters 3 and 5.

However, since the early 1970s there has been an increasing and rapidly accelerating move away from acidic systems towards neutral and even slightly alkaline pH. The advantages of operating at higher pH are that there is reduced corrosion, greater strength arising from better swelling of fibres at higher pH, the possibility of using high filler additions, and the energy savings associated with the easier drying of filled paper. This change has had a profound effect upon the whole of wet-end chemistry.

A further complication is that modern papermaking often involves closed white water systems, fast machines and twin-wire forming. These all require new approaches to the chemistry of the wet end, and this text attempts to summarise the current understanding of modern wet-end chemistry in terms of the basic principles which underpin it.

The variety of products which are available on the market becomes ever more complicated, but a summary, which is regularly updated, is available elsewhere [2].

Table 1.1 Pulp paper and board per capita annual consumption 1983–1989 (kg)

Industrial area	Annual consumption per capita						
	1983	1984	1985	1986	1987	1988	1989
Nordic	180	192	224	232	205	244	234
Western Europe	120	129	129	135	135	152	157
USA	267	287	284	290	291	311	304
Japan	153	169	168	173	173	204	223

1.2 Paper chemical use in specific product grades

The size of the paper chemicals market is, as would be expected, related to the size of the paper and board manufacturing sector and also to the distribution of product types within the industry. It is therefore instructive to look at the international paper and board market in the industrially developed world. Table 1.1 shows the consumption per capita of paper and board products over the period 1983 to 1989 for several key industrial areas [3].

The average annual growth per annum for these countries during this period was 4.6%. However, this increase conceals some fairly large variations in individual product growth. Table 1.2 shows the total production for USA and Western Europe of various grades of paper and board products for the period 1983–1986.

The strongest growth has been in newsprint and in coated and uncoated printing and writing grades, whilst tissue and board have shown steady but less strong growth. These figures have important implications for the paper chemicals market and it is worth considering these implications for the various product types shown in Table 1.2.

Table 1.2 Total annual production (millions of tonnes) of different grades of paper and board for USA and Western Europe during the period 1983 to 1986

Grade	Total annual production				Mean % annual growth
	1983	1984	1985	1986	
Newsprint	10.3	11.4	11.5	11.9	4.8
Coated printing & writing	10.3	11.8	11.5	12.3	6.3
Uncoated printing & writing	19.6	20.9	21.4	22.4	4.5
Tissue & sanitary	6.6	6.8	6.8	7.1	2.5
Packaging & industrial	8.4	8.5	8.0	7.8	−2.1
Board	43.8	46.5	45.5	48.2	3.3

1.2.1 Newsprint

Newsprint production has traditionally involved only minimal use of chemical additives. The product is usually unsized and, because of its low cost, the use of expensive retention aids is not considered to be cost-effective. However, this situation will probably change as a result of the trend towards limited filler inclusion and also the increased use of deinked waste paper in the furnish. The use of retention aids is, however, likely to be cost-effective for environmental reasons rather than for reasons of raw material conservation.

Deinking of recycled waste paper is performed either by flotation or by simply washing the fibres with dispersants or surfactants or by a combination of both. In these processes, the ink particles are released from the fibre matrix by a dispersant and are then either washed from the pulp or floated by rendering the particles hydrophobic and attaching them via a collector molecule to air bubbles. The ink-rich bubbles float to the surface as a foam which is then removed. The growth in deinking chemicals has paralleled the growth in the use of deinked fibre in newsprint. The surface chemistry of flotation and wash deinking is poorly understood and, at the moment, could be more accurately described as an art rather than a science. Much more basic work is required to improve our understanding of it. The subject has not been dealt with in this book as it is more strictly a part of pulping and recycling chemistry than wet-end chemistry. It is also covered in other texts [4,5].

The production of newsprint is generally an acidic process due to the naturally low pH of groundwood pulp. Conversion to alkaline conditions is not therefore likely to take place very readily in the short term, although this situation may change as calcium carbonate is found increasingly in recycled fibre.

1.2.2 Printing and writing grades

One of the strongest growth areas in the past few years has been in the area of printing and writing grades, both coated and uncoated. Much of this increase in demand can be accounted for by commercial printing and advertising (often lightweight coated), and by computer print-out and copier paper. Many of these printing and writing grades are now made under neutral or slightly alkaline conditions, whilst many board grades have been made at these pHs for some time. Conversion to alkaline systems for coated grades has been particularly pronounced because of the demand for higher quality and because of the compatibility of calcium carbonate-filled sheets with calcium carbonate-based coating formulations.

The trend to alkaline pHs in these products has very important implications for paper chemicals. The sizing system must be changed from rosin-alum to a neutral system (usually AKD or ASA), and this is discussed more fully in Chapter 9. Retention aids appropriate to alkaline systems are also required, and the choice of wet-strength agent is also dependent upon pH. There is also a greater tendency for pitch and slime deposits to develop in neutral or alkaline systems and this necessitates the increased use of biocides and pitch controlling chemicals. The shift to higher pH has also increased the use of dry-strength agents, particularly cationic starch and polyacrylamides. This is because they help to offset the strength losses associated with highly filled paper.

There are some clear effects in the market for chemicals in coated grades, in particular in terms of the coating formulations. There has been a move towards the use of synthetic polymers in these formulations (e.g. acrylates, styrene-butadiene copolymers and polyvinyl alcohols and acetates) rather than natural polymers and their derivatives (e.g. carboxymethyl cellulose and starch). This is because of the better quality of such formulations for glossy advertising paper. However, coating is not a subject which is covered in this book and the reader is referred to other texts [6,7].

1.2.3 Tissue and sanitary

Probably the biggest change in recent years has been the trend towards the increased use of recycled fibre in tissue manufacture. This is likely to continue as consumer pressure for recycled products of this type grows, and deinking will continue to grow in importance in this product area. Tissue softening is not a subject which has been dealt with specifically in this book but it is an important area of chemical application and will probably grow in importance as the use of recycled fibre, which produces less soft tissue, increases. Softening agents vary greatly in chemical type and mechanism of action. Debonding agents, in which a cationic water soluble hydrophobic molecule is bound to the fibre and thus interferes with bonding, are popular but there are also those which change the textural quality of the sheet. The most common debonding agents in the former category are the quaternary ammonium type, with hydrophobic, long chain, alkyl substituent groups.

The increased use of recycled waste paper will also necessitate the increased use of biocides and pitch controlling agents to improve runnability. The use of dry-strength agents is also likely to increase.

1.2.4 Packaging and board

Many packaging and board grades are made from recycled fibre, particularly in Europe. The chemical technology of recycling and deinking will therefore continue to be of importance in these products, and the control of pitch and deposits is essential. Many grades require both wet- and dry-strength properties. There has been an increasing tendency to use alkaline systems and, hence, synthetic sizing for the production of many packaging grade boards, for example aseptic liquid packaging board for end-uses such as juice cartons and fast food containers. Most liner board is now also made under alkaline conditions. Sizing characteristics are often very important in these products. In particular the rate at which sizing develops is often a key requirement (see Chapter 9). In the future, new chemicals are likely to be developed as a result of the more stringent criteria which will probably be applied to chemical additives used in food packaging grades.

1.3   Trends in paper chemical use

Paper chemicals, if they are to be effective, must usually be retained in the sheet. This is essentially a physical and colloid chemistry problem. The particles which comprise the suspension of fibres, fines and filler – and which therefore constitute the papermaking system – exhibit, at normal pHs and in the absence of chemical additives, a negative charge (anionic). Many chemical additives, in order to given them a high affinity for the furnish components, are therefore positively charged (cationic). The interaction between these charged species, and in particular the influence of the aqueous environment on this interaction, are of crucial importance in determining retention. The highly anionic nature of the fines allows them to compete with the fibre and filler surfaces for cationic species, and their presence therefore is an important variable in the retention of cationic paper chemicals. The determination of the charge demand of papermaking systems is therefore of great importance and is discussed in Chapters 3 and 5.

Even though the system is inherently anionic it is possible, under some circumstances, to use anionic additives. This subject is also discussed more thoroughly in various chapters in this book.

1.3.1   *Deinking chemicals*

Deinking is performed at alkaline pHs. In the wash deinking process, a dispersant or surfactant is used to separate the ink particles from the fibre. These are most commonly alkyl phenol ethoxylates, but, to a lesser extent, polyphosphates are also used. In flotation deinking, saponified fatty acids in the presence of calcium ions are the major chemical components.

In addition to these chemicals, deinking may involve the use of bleaches such as hydrogen peroxide or sodium hypochlorite. In the case of the former, sodium silicate is often necessary for stabilisation purposes.

1.3.2   *Defoamers*

Most paper mills make use of defoamers to prevent excessive foaming during the approach flow system and during sheet formation. They may also be used in coating operations. They are surface active agents such as polyethylene glycols, and esters or amides of fatty acids.

1.3.3 *Retention and drainage aids*

In recent years, growth in the use of retention aids has been greater than that of almost any other paper chemical additive. It has been caused by a combination of factors: increased machine speeds, the increased use of fillers in alkaline systems, the increased use of recycled paper and the growing tendency to use fillers in newsprint.

Cationic, neutral and anionic polymers are used either singly or in combination (Chapter 5) but the trend has been towards the increased use of cationic polyelectrolytes. The polyacrylamides and polyethyleneimines are the market leaders in this group but colloidal silica is becoming increasingly important. Cationic starch is an important component of colloidal silica dispersions, and it also contributes to dry strength. The polyacrylamides are also able to behave as dry-strength additives (Chapter 6). This pattern is part of an increasing trend towards multicomponent chemical addition.

1.3.4 *Dry-strength additives*

Starch, because of its competitive price, is overwhelmingly the most important dry-strength additive used in papermaking today. About twenty times more starch than polyacrylamide is used for dry strength. The starches for wet-end application are now almost always cationic and therefore, to some extent, also function as drainage and retention aids. They are also used to stabilise size dispersions (Chapter 9).

1.3.5 *Sizing*

Sizing chemistry is inexorably linked to the pH of the papermaking system, and the trend towards neutral papermaking has meant that there has been a steady decline in the use of rosin and alum (see Chapter 8) and a concomitant increase in the use of directly reactive sizes such as alkenyl succinic anhydrides (ASA) and alkyl ketene dimers (AKD) which operate more effectively at high pH (Chapter 9). This trend has been more pronounced in Europe than in Scandinavia and North America, for reasons which are discussed in Chapter 9. In Japan, rosin–alum is still the dominant sizing process, but the use of AKD and ASA (dominant) is increasing.

In North America, the AKDs are used slightly more than the ASAs, but the greater growth in the past few years has been in ASA. In Europe, the AKDs are dominant and there is very little use of ASA. However, this situation will probably change over the next few years.

1.3.6   Wet-strength additives

Wet-strength additives (Chapter 7) are of major importance in many paper and board product grades, such as packaging papers, sacking, poster paper, etc. The main products used are urea–formaldehyde (U/F) and melamine–formaldehyde (M/F), which tend to be used in acid systems, and polyamine–polyamide–epichlorohydrin resins which are more suited to neutral and alkaline systems.

Opportunities exist for the development of more environmentally acceptable products. The U/F and M/F resins always contain traces of free formaldehyde, and the organo-chlorine content of epichlorohydrin resins gives rise to measurable absorbable organic halogen (AOX) in products and water systems.

1.3.7   Wet-end chemistry control

Probably the most active area of interest in wet-end chemistry over recent years has been in the attempts to create a control strategy for wet-end chemical addition. The greatest problem here is that it is not yet, nor is likely to be, possible to generate a comprehensive physico-chemical model for the description of the adsorption, retention and other processes operative at the wet end of a multi-component additive system. However, some success in control has been achieved with more empirical approaches. The trend has been to monitor as closely as possible those wet-end chemistry variables that are amenable to accurate measurement. Improved instrumentation has made this possible in a number of areas, particularly retention and charge demand. These techniques may then enable statistical control strategies or expert systems to be introduced for the control of the chemistry of the wet end. The subject is more fully discussed in Chapter 12.

References

1. Lindström, T., *Trans. 9th Fundamental Research Symp.*, Cambridge, England, eds, Baker, C.F. and Punton, V.M., Mech Eng. Publ. Ltd, London (1989), 311–435.
2. Schultz-Edwards, O. ed., *Winterchem Catalogue*, Benn Brothers, Germany (1990).
3. Source: British Paper and Board Industry Federation.
4. Larsson, A. *et al.*, 'Surface chemistry of the deinking process. Part 3: Deposition of ink and calcium soap particles on fibres', *Svensk Pappersdidn.*, **88** (3) (1984), R1-R7.
5. Ortner, H.E., *Recycling of Papermaking Fibres – Flotation Deinking*. TAPPI monograph. TAPPI Press, Atlanta (1981).
6. Garey, C.L. (ed.), *Physical Chemistry of Pigments in Paper Coating*, TAPPI Press, Atlanta (1977).
7. Casey, J.P. (ed.), *Pulp and Paper Chemistry and Chemical Technology*, Vol. IV, J. Wiley and Sons (1983).

2 Accessibility of cellulose

G.A.F. ROBERTS

2.1 Introduction

Cellulose, in common with many other polymers, has a two-phase morphology containing both crystalline and non-crystalline (amorphous) material. The model most frequently used is the fringed micelle structure [1] in which individual chains pass through several crystalline and non-crystalline regions (Figure 2.1). Although this model is in agreement with many of the properties of cellulose [2] other models have been proposed, in many instances based on the observed properties of the microfibrils. These include the fringed-fibril structure [3] and paracrystalline structures [4–11] in which the non-crystalline component arises from dislocations and chain-ends within the highly crystalline fibrils [12] making up the microfibril (Figure 2.2). Krässig [13] has proposed a further model in which the elementary crystalline fibril is surrounded by one layer of amorphous ('disturbed') chains, whilst the layer of chains immediately within this amorphous layer is accessible to reagents such as water or deuterium oxide. Despite the occurrence of chain folding in cellulose single crystals [14] most of the available evidence is against such an arrangement in native cellulose [15–19].

2.2 The concept of accessibility

Although the crystalline regions may be quite clearly defined, for example in the definition used by Mann and Marrinan [20], as being those regions possessing a three-dimensional order sufficient to give a regular repeating system of hydrogen bonds, it is more difficult to define the non-crystalline regions. This is due to the wide range of levels of disorder between the two extremes of crystalline and complete disorder. Howsmon and Sisson [21] depicted the gradual transition between these two extremes for a volume element (Figure 2.3). The degree of order ($\bar{O}$) is defined by the equation

$$\bar{O} = [OH_c]/[OH_t]$$

where $[OH_c]$ is the total number of hydrogen bonds in the region and $[OH_t]$ is the total number of hydrogen bonds possible if all the chains are perfectly

Figure 2.1   Fringed micelle model of two-phase polymers.

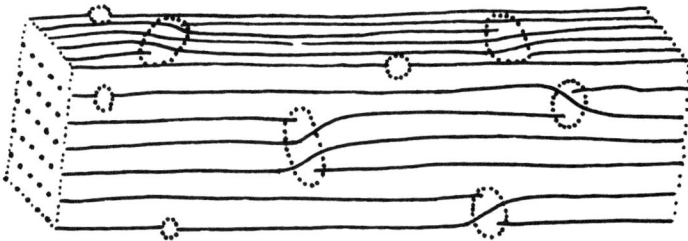

Figure 2.2   Structural model of elementary fibrils as proposed by Mühlethaler [9].

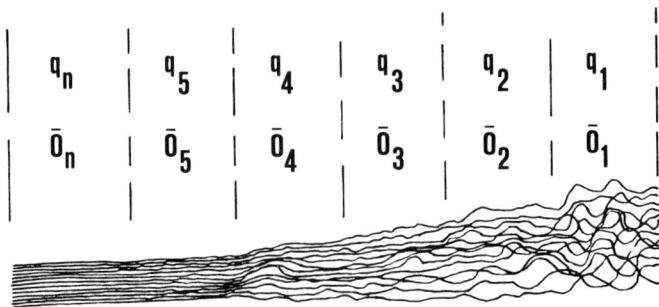

Figure 2.3   Schematic mass-order distribution in cellulose, after Howsmon and Sisson [21].

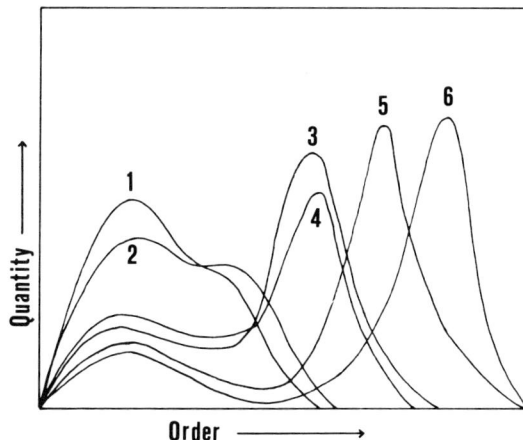

Figure 2.4 Typical lateral-order distribution curves for different cellulose fibres after Howsmon and Sisson [21]. Curves: 1, high-tenacity rayon; 2, textile rayon; 3, high-strength, low-elongation rayons; 4, mercerised native fibres; 5, wood pulps; 6, cotton.

crystallised. If q_i represents the fraction of total cellulose having a degree of order of $\bar{O}_i$, then

$$Q_n = \sum_{i=o}^{i=n} q_i$$

where Q_n is the amount of cellulose present in a sample having a degree of order up to and including $\bar{O}_n$. Information regarding values for the different levels of order present in a cellulose sample enable lateral-order distribution curves to be plotted (Figure 2.4).

Since most chemical reagents, including water, penetrate only the non-crystalline regions, the existence of regions having differing degrees of molecular disorder means that the extent to which a reagent can penetrate the cellulose structure will depend both upon the molecular size of the reagent, and upon the extent to which the cellulose is swollen under the experimental conditions. As Howsmon and Sisson pointed out [21], any experimental technique which is used to determine the percent crystalline or the percent non-crystalline content of a cellulose sample defines two quantities

$$Q_a = \sum_{i=o}^{i=k} q_i \quad \text{and} \quad Q_c = \sum_{i=k}^{i=n} q_i$$

where Q_a and Q_c are the non-crystalline and crystalline fractions respectively. However, the application of any two techniques will only give results that are in agreement if the two methods differentiate between crystalline and non-crystalline cellulose at the same order level. Since this is normally not the case, it is more realistic to talk in terms of non-accessible and accessible regions, the volume ratio of which will depend upon the particular method used for its determination.

In addition, the non-accessible fraction cannot be equated directly with the crystalline fraction for two reasons. Firstly, defects within the crystalline regions may not be accessible to any of the reagents used. Secondly, the hydroxyl groups on the surfaces of the crystalline regions are accessible to some reagents, hence the accessible fractions of two samples having the same crystallinities may vary if the size and shape of the crystalline regions vary. An equation relating accessibility and crystallinity has been proposed by Frilette *et al.* [22]:

$$A = \sigma\alpha + (100 - \alpha)$$

where A = percent accessible cellulose,
  α = percent crystalline cellulose,
  σ = fraction of crystalline regions accessible to the reagent in question.

If both A and α are determined independently it is possible to calculate σ and this enables an estimate of the surface areas of the crystalline regions to be made [23]. However, the values obtained should be treated with considerable caution as the 'accessible regions' determined may exclude some higher order levels that are not completely crystalline.

In the opinion of Tripp [24], accessibility is a more important parameter than crystallinity: '... the methods that measure accessibility to actual or model reagent systems may be more pertinent criteria of properties of technological importance than the procedures that estimate overall three-dimensional order'. Certainly the accessibility of the non-crystalline fraction under conditions of use is of fundamental importance to the chemical reactivity, sorptivity, permeability, retention of dyes and sizing materials, pulp drainage and fibre-fibre bonding behaviour. This point is also taken up in later chapters. Complete characterisation of the cellulose structure requires, in addition to the molecular arrangement within the crystalline regions, a knowledge of the relative amounts of accessible and non-accessible material, the size and shape of the crystalline regions and the order distribution of the non-crystalline regions.

2.3   Determination of crystallinity/accessibility of cellulose

Physical, chemical and sorption methods have been used for these determinations.

Physical methods, which are based on the fact that crystalline cellulose has different physical properties to those of non-crystalline cellulose, include X-ray diffraction, density determination and IR spectroscopy. Chemical methods may be sub-divided into methods carried out under swelling

conditions, including acid hydrolysis, deuterium exchange, aqueous formylation and periodate oxidation, and those carried out under non-swelling
conditions, for example chromic acid oxidation, thallation and vapour phase
reaction with formaldehyde. Sorption methods may be similarly sub-divided
into moisture sorption, iodine sorption and bromine sorption, taking place
under swelling conditions, and nitrogen sorption which takes place under non-
swelling conditions. In general the physical methods determine the percent
crystallinity whilst the chemical and sorption methods determine the
accessibility.

2.3.1   Physical methods

2.3.1.1   X-ray diffraction. A typical X-ray diffractogram for native cellulose is
shown in Figure 2.5. The maxima are produced by specific reflections from the
crystalline regions, the number assigned to each maximum indicating the
lattice plane giving rise to it. In order to determine the percent crystallinity of
the sample it is necessary to separate these specific reflections from the more
diffuse scattering due to the non-crystalline regions, and a number of methods
for achieving this have been proposed. However, it should be borne in mind
that Mann [25] has argued that estimates of crystallinity made from such
diagrams have no absolute significance and that the significance of even
relative crystallinities is uncertain.
    The first quantitative method [26, 27] involves comparison of the integrated

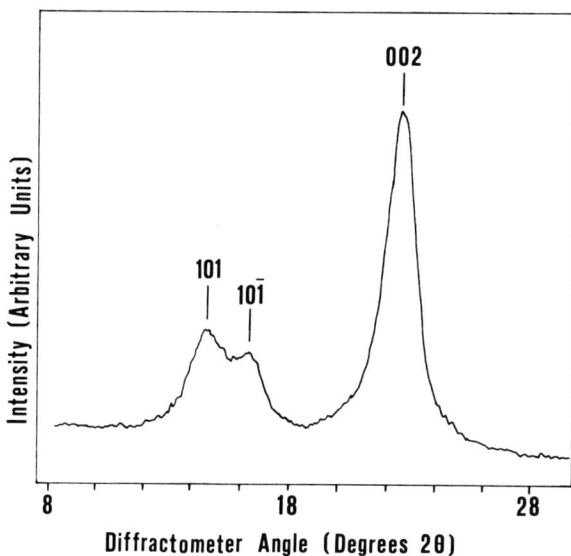

Figure 2.5   Typical X-ray diffractogram for cellulose I (cotton).

intensity of the maxima and that of the diffuse scattering. The separation into crystalline and non-crystalline scattering has always been a somewhat arbitrary procedure [28] although there is an agreed method [25]. More recent developments in this approach have involved extending the range of the measured angle of diffraction to approximately $2\theta = 60°$ [29,30]. However, very small differences in the percentage crystallinity were obtained for a wide variety of samples; indeed Shenouda and Viswanathan [30] found that all the cotton and viscose rayon samples they examined had 30–32% crystallinity. It has been suggested by Tripp [24] that this extension of the scattering range to $2\theta = 60°$ adds an area over which a strong background persists and in which relatively small, discrete reflections are present, and that this would compress the range of values obtained.

The second approach involves the use of crystalline and amorphous standards, those chosen being acid hydrolysed cellulose and ball-milled cellulose representing 100% crystalline cellulose and 100% amorphous cellulose respectively [31]. If U, C and A are the corrected scattering powers of the sample, crystalline standard and amorphous standard respectively at a given value of 2θ, then $U = fC + (1 - f)A$ where f is the degree of crystallinity. Thus a plot of $(U-A)$ versus $(C-A)$ for corresponding values of 2θ gives a straight line passing through the origin, the slope of which is the degree of crystallinity. The validity of this method depends on the correctness of the standards chosen. Although prolonged ball-milling of cellulose results in almost complete disappearance of the characteristic X-ray diffraction maxima [32] it is not so certain that hydrolysis produces 100% crystalline material, crystal defects within the crystalline regions being inaccessible to the acid. The method has been extended to cover samples containing more than one crystalline phase [33].

The third approach involves relating the intensity of one of the principal diffraction maxima and that of an appropriate minimum in the diffractogram to give a 'crystallinity index' [34–36]. Segal et al. [35] selected the intensity of the 002 maximum at $2\theta = 22°$ (I_{002}) and that of the minimum at $2\theta = 18°$ (I_{am}) as the two points and defined the crystallinity index as

$$CrI = [(I_{002} - I_{am})/I_{002}] \times 100$$

Good correlation was found between the crystallinity index values and the percent crystallinity obtained by acid hydrolysis (see later), the correlation coefficient being 0.98. However, the crystallinity index values gave a rather compressed scale ranging from 79.0–64.9% for samples whose percent crystallinities, as measured by acid hydrolysis, varied from 88.6% to 24.0%. This latter sample had a non-accessible content of 38% as determined by moisture regain (see later).

A related approach has been developed by Ellefsan et al. [37] using ball-milled cellulose as a standard for 100% non-crystalline cellulose to determine the ratio of the intensity of the minimum at $2\theta = 18°$ for the sample to that for the ball-milled standard.

An excellent review of the application of X-ray diffraction techniques in the determination of the crystalline content of cellulose has been given by Tripp [24].

2.3.1.2 *Density measurements.* This method involves the determination of the macroscopic density in a suitable non-penetrating liquid, either by the method of Hermans [38] or by use of a density gradient column [39]. From this, the ratio of crystalline to non-crystalline material may be calculated from the respective densities of the two phases. That for crystalline cellulose may be calculated from the unit cell volume, but that for the non-crystalline component cannot. Hermans [40] assumed the difference between the densities for the two phases to be similar to that between those of crystalline and amorphous butyl alcohol, namely 6%. A major objection, apart from the fact that no justification was given for selecting butyl alcohol for comparison, is that since there is a range of degrees of disorder in the non-crystalline regions and their distribution may vary between celluloses of different origins, the concept of a specific value for the density of the non-crystalline material is not valid.

Other workers [35,41] have accepted the basic concept of Hermans that the macroscopic density will differ with variation in its crystalline content, but have used the density itself as an index of crystallinity rather than attempt to calculate the relative proportions.

2.3.1.3 *Infrared spectroscopy.* Although there would appear to be no absorption band in the IR spectrum of cellulose that can be related directly to the crystalline content (with the exception of the O–H [st] in the 3600–3400 cm^{-1} region – see later), the bands at 1429 cm^{-1} (CH_2 scissoring) and 1372 cm^{-1} (CH bending) decrease in intensity with any decrease in crystalline content, whilst that at 893 cm^{-1} (C_1 group vibration) increases. The band at 2900 cm^{-1} (CH stretching) appears to be independent of the crystalline content of the sample. Nelson and O'Connor [42] proposed the use of the ratio A_{1429}/A_{893} as an IR crystallinity index and demonstrated that the value of the index varied from 2.80 for untreated cotton down to 0.11 for ball-milled cellulose. They subsequently introduced a second crystallinity index based on the ratio A_{1372}/A_{2900}; this ratio may be used with samples containing a mixed lattice, whereas the A_{1429}/A_{893} ratio cannot [43,44]. Although the technique is convenient and requires only very small amounts of sample, it is necessary to standardise the operating conditions and sample presentation very carefully if good reproducibility is to be achieved [24].

2.3.2 *Chemical methods*

2.3.2.1 *Acid hydrolysis.* The method is based on the premise that chemical reactions proceed much more rapidly in the non-crystalline than in the crystalline regions so that hydrolysis rate curves may be interpreted as the superposition of the rapid reaction in the accessible regions on the much slower reaction in the crystalline regions. The rate of hydrolysis may be followed by catalytic oxidation of the D-glucose produced and determination of the CO_2 evolved [45–47] or by determination of the unhydrolysed cellulose residue [25, 35, 48, 96]. This latter method is preferable but in either case extrapolation of the rectilinear portion, which represents the slower hydrolysis of cellulose chains in the crystalline regions, back to zero time gives a measure of the crystalline content of the sample. The results obtained will tend to indicate too high a crystalline content, due to crystallisation of some of the chains in the accessible regions during the hydrolysis step [23].

2.3.2.2 *Periodate oxidation.* This method is also based on the assumption that the reaction between periodate ions and cellulose, which has been comprehensively reviewed [49,50] occurs much more rapidly in the accessible regions [51,52] and that the rate curves may be analysed in a similar manner. Later workers showed that rectilinear plots suitable for extrapolation to zero time are more readily obtained by the first order plot of log C versus time, where $C = \%$ unreacted oxidant, than by plotting oxidant consumed versus time [53].

2.3.2.3 *Formylation.* This method depends on the observation that, under well-defined reaction conditions, formic acid reacts only with the primary hydroxyl groups of the anhydro-D-glucose residues and hence the extent of formylation is a measure of the accessibility of the cellulose sample [54–57]. That the incomplete reaction of the primary hydroxyl groups in cellulose samples is due to lack of accessibility, rather than to the setting up of an equilibrium system, has been clearly demonstrated by the use of starch as a representative 100% accessible material, when 100% formylation of the primary hydroxyls occurs [54].

2.3.2.4 *Deuterium exchange.* On interaction of cellulose with D_2O, either as a vapour or as a liquid, the accessible hydroxyl groups undergo deuterium exchange. Various methods have been used to follow the reaction and determine the extent of exchange. These include density [22], refractive index [20], mass spectrometry [58,59] and IR spectroscopic measurements [41,60,61] on the liquid phase, and weight increase [7] and IR spectroscopic measurements [20,62–65] on the deuterated cellulose itself. Frilette *et al.* [22] found that there is a rapid initial reaction which is completed within 4 hours, after which very little change occurs over the next 168 hours, and they equated the fraction unreacted after 4 hours with the crystalline fraction. This conclusion was supported by other workers [20,66,67] and conflicting results obtained using IR spectroscopy [62] have been attributed to re-hydrogenation of samples by atmospheric water vapour, prior to the IR measurements.

The most detailed studies of deuterium exchange are those of Mann and Marrinan [20,63]. These workers showed that the exchange could be limited to the non-crystalline regions if deuteration is carried out using D_2O vapour rather than liquid, and used the ratio of the absorbances of the OH band at $3450\,cm^{-1}$ (approx.) and the OD band at $2550\,cm^{-1}$ (approx.) to produce a relative scale of crystallinity values for a series of cellulose samples. These were converted to an absolute scale by extracting an exchanged sample with H_2O, after measuring the A_{OD}/A_{OH} ratio, and determining the quantity of deuterium extracted by refractive index measurements of the H_2O/D_2O mixture. The technique, as applied by them, was better suited to cellulose in film form than in fibre form but more recently the application of the technique to fibres has been reported [65].

2.3.2.5 *Chromium trioxide oxidation.* Chromium trioxide in acetic acid/acetic anhydride rapidly oxidises cellulose chains in accessible portions and interacts much more slowly with the crystalline fraction [68,69]. Thus the rate of reaction curves may be treated similarly to those obtained for periodate oxidation, and the accessible fraction calculated. The non-swelling nature of the reaction medium was demonstrated by Gladding and Purves [68] who found that a dry, swollen cotton sample underwent considerable oxidation, whilst a dry, unswollen sample underwent minimal oxidation, as did a sample of dry, powdered starch.

2.3.2.6 *Reaction with thallous ethylate.* Cellulose undergoes thallation of accessible hydroxyl groups on reaction with thallous ethylate in benzene. Quantitative replacement of thallium by methyl groups occurs on reaction with excess methyl iodide and subsequent determination of the methoxyl content gives a measure of the accessible hydroxyl groups in the original sample [70–72]. Again the reaction takes place in a non-swelling system.

2.3.2.7 *Vapour phase reaction with formaldehyde.* The extent of reaction of cellulose with formaldehyde vapour, compared with that of dextrin which was taken as a reference material having 100% accessibility, has been used to determine the accessibility of cotton. However, the conditions used, no liquid phase and a reaction temperature of 125°C, are so different from those used for the other chemical reaction methods that it is not surprising that there is only one report of its use [73].

2.3.3 *Sorption methods*

2.3.3.1 *Iodine sorption.* The test procedure as originally proposed by Schwertassek for determining relative accessibilities [74] is subject to a number of experimental variables which must be carefully standardised if reproducible results are to be obtained [75,76]. Furthermore, although it has been suggested that the results may be interpreted in terms of monolayer adsorption and hence may be used to calculate internal surface areas [77], this has been criticised by Wadsworth and Cuculo [78] and the method remains essentially one which provides an index of the accessible content through iodine sorption values (ISV). Only in a very few cases have ISV figures been converted to percent accessible material values [79, 44].

2.3.3.2 *Bromine sorption*. Bromine sorption, which has been extensively studied by Lewin and co-workers [80–82], has a number of advantages over iodine sorption: the technique does not cause swelling in the sample and, provided correct conditions of solution concentration and liquor ratio are used, it is reproducible and reasonably rapid. However, the main advantage over iodine sorption is that the Br_2 molecules are monomolecularly adsorbed onto the glycosidic oxygens. This enables the percent accessible material to be calculated from the saturation uptake value obtained from a Langmuir plot of the equilibrium values.

2.3.3.3 *Moisture sorption*. Although several workers had previously studied the sorption of moisture by cellulose, Hermans [83] was the first to suggest a direct relationship between the moisture regain of a fibre and its content of accessible material. Prior to this, Urquhart and Williams [84] had introduced the concept of the sorption ratio, defining it as the ratio of the moisture regain of a cellulose sample to that of pure cotton measured under the same conditions of temperature and relative humidity. Hermans [40] and Howsmon [23] subsequently proposed that the sorption ratio, which is constant over the relative humidity range of 10–70% (approximately) should provide a relative measure of the accessible content of a cellulose sample. Both workers also suggested that the product of the percent accessible content of native cotton and the sorption ratio of a cellulose sample gives the percent accessible material of that sample.

By extrapolation of moisture regain data for randomly substituted methyl celluloses and cellulose acetates, Gibbons [85] determined that, at 25°C and 65% relative humidity, fully accessible cellulose adsorbs 1.53 molecules of water per anhydro-D-glucose unit (AGU). This is in very good agreement with the value of 1.51 molecules H_2O/AGU determined later for a ball-milled cellulose sample having a 98% accessible content as determined by deuterium exchange [86]. The former value (1.53 molecules H_2O/AGU) is equivalent to a moisture regain of 17.0% and so the accessible content may be obtained directly from the moisture regain value [75,87].

2.3.3.4 *Nitrogen sorption*. Nitrogen sorption measurements, and application of the BET equation, enable the internal surface areas of cellulose samples to be determined [71,88–91]. Although no direct conversion between

Table 2.1 Typical percent accessibility values obtained using different techniques

	Substrate			
Technique	Cotton	Wood pulp	Mercerised cotton	Regenerated celluloses
Physical				
Density	36	50	64	65
X-ray diffraction	27	40	49	65
Chemical				
Acid hydrolysis	12	17	24	32
Deuterium exchange	42	55	59	68
Formylation	21	31	35	65
Periodate oxidation	8	8	10	20
Chromium trioxide oxidation	0.3			
Thallation	0.85			
Sorption				
Bromine	21			31
Iodine	13	23	32	40
Moisture*	40	48	62	74
Moisture†	42	52		

*Calculated from sorption ratio assuming a value of 40% for the accessible content of cotton.
†Calculated from moisture regain assuming a regain value of 17% for 100% accessible cellulose.

surface area from nitrogen sorption and accessibility, as measured by any of the above techniques, has been established, both types of measurement follow the same trends with variation in substrate structure.

2.3.4 *Comparison of results from different techniques*

Typical results from the application of a number of the above techniques for up to four different cellulose substrates are given in Table 2.1. Despite the considerable range of accessibility values for any substrate there is excellent agreement between the different techniques in placing the substrates in order of increasing accessibility:

cotton < wood pulps < mercerised cotton < regenerated celluloses

Generally deuteration and moisture regain measurements give the largest values for the accessible fraction. This is understandable in view of the considerable swelling effect of water on cellulose and the fact that the surfaces of the crystalline regions are accessible to H_2O and D_2O. Although bromine sorption also takes place from an aqueous medium, the crystalline surfaces are not available to the sorbate since the glycosidic oxygens of the outer chains are not on the surface but are embedded within the crystalline region to a depth of half the chain width. Hence lower accessibilities would be expected. The low values from acid hydrolysis studies are due to crystallisation of chain segments produced on hydrolysis of chains initially in the accessible regions and to the inaccessibility of the glycosidic oxygen atoms of the chains on the surfaces of the crystalline regions.

The very low values obtained with chromium trioxide oxidation and with thallation are due to the non-swelling nature of the reaction media used in these methods. Formylation, in which the reaction medium is formic acid, gives a much greater value for the accessible fraction, due to the greater swelling power of formic acid compared with glacial acetic acid/acetic anhydride or benzene.

Not surprisingly, good correlation has been obtained between moisture regain values, or sorption ratios, and accessibilities determined by deuterium exchange [23,59,61,92] and more specifically between the latter, when the extent of deuteration is determined by density measurements or IR spectroscopy, and accessibilities calculated from moisture regain values [23,75]. In these cases the actual values for accessibility obtained by the two methods correspond, but when the deuteration reaction is measured by mass spectrometry [59] this is not the case.

Moisture regain results have also been found to be correlated with accessibilities determined by acid hydrolysis and periodate oxidation [75], although the accessibilities from moisture regain were consistently higher, and with the IR crystallinity index [42, 44, 78].

Good correlation has been found between the X-ray crystallinity index and moisture regain [35] and between the former and the IR crystallinity index [35,43]. However, only moderate correlation was found between the non-crystalline fraction calculated from X-ray diffractograms [26, 27] and moisture regain [41].

Nitrogen sorption and thallation have been used to quantify the extent of swelling brought about by different treatments, particularly swelling in water and NaOH solution. As drying the substrate after swelling in water reduces the accessibility to close to the initial value, the swollen cellulose is dried by a solvent exchange process [93] followed by analysis under the non-swelling conditions of the two techniques. Typical results are given in Table 2.2 and demonstrate the dramatic increase in accessibility and/or surface area brought about by immersion in water or NaOH solution. It is of interest that the surface area of unswollen cotton fibre, as measured by nitrogen sorption, is close to that calculated by Howell and Jackson [94] from microscopy measurements, indicating that in this case nitrogen sorption is measuring the macroscopic surface and that penetration occurs to a very limited extent.

These techniques are based on the assumption that, during the solvent exchange process, there is no change in structure compared to that in the swollen state. However, Merchant [90] and Sommers [91] have shown that the value of the surface area obtained can be influenced both by the intermediate solvent and by the final, non-polar, solvent used in the exchange

Table 2.2 Typical values of accessibilities and surface areas for swollen, solvent exchanged, cotton

Substrate	Accessibility (%)	Surface areas ($m^2 g^{-1}$)	
	Thallation	Thallation	Nitrogen sorption
Cotton fibre	0.85	16	0.55
Cotton fibre, swollen in water	13.6	263	137
Cotton fibre, swollen in 2.5 M NaOH	10.1	195	71.3
Cotton fibre, swollen in 5 M NaOH	12.4	240	–

process. Nevertheless the results confirm the considerable increase in accessibility/surface area brought about by water.

Similarly large surface areas of up to $140 \, m^2 g^{-1}$ have been obtained by Sommers [91] from analysis of water vapour sorption isotherms on purified cotton. No prior swelling treatment was given to the samples, swelling taking place on adsorption of water vapour.

The majority of the work on accessibility determination has been carried out using cotton as substrate because it is the most consistent source of cellulose and is readily purified. In the case of wood pulps the processing treatments may cause some chemical modification, principally the formation

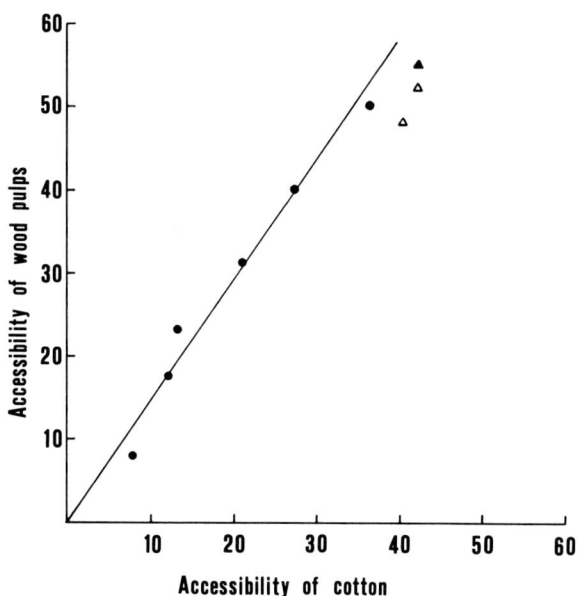

Figure 2.6 Plot of the percent accessibility values for wood pulps, measured by several techniques, versus the values for cotton measured by the same techniques. Values are taken from Table 2.1. ▲ Deuterium exchange; △ moisture regain; ● remaining techniques.

of carboxylic acid groups, and also increased fibrillation of the fibre. The presence of carboxylic acid groups would be expected to increase the degree of swelling, through both osmotic effects and the affinity of the carboxylic acid group for water. Klug [95] has shown, using a range of water-soluble cellulose ethers, that the relationship between equilibrium moisture regain and the chemical composition (C : O ratio) is linear for non-ionic derivatives, whilst the regain value for CMC is much greater than is predicted from its composition.

Since any chemical modification will take place in the accessible regions of the fibre, this increase in swelling should not result in an increase in the accessibility value. This is demonstrated in Figure 2.6 where the typical percent accessibility values for wood pulps, given in Table 2.1, are plotted against the values for cotton using the same techniques. It can be seen that there is approximately the same ratio between the values obtained for the two substrates by any of the techniques, whether physical or chemical. However the results for moisture sorption and deuterium exchange fall below the line (Figure 2.6). This indicates that the accessibility of wood pulp as measured by these techniques is less, relative to that of cotton, than would be expected from consideration of the results from the other techniques. This is surprising since the introduction of COOH groups during pulping should increase the hydrophilic character of the pulps and hence increase the accessibility values when determined by moisture sorption.

References

1. Kratky, O., *Kolloid-Zeit.*, **70** (1935), 14–19.
2. Scallan, A.M., *Text-Res. J.*, **41** (1971), 647–653.
3. Hearle, J.W.S., *J. Pol. Sci.*, **28** (1958), 432–435.
4. Rånby, B.G., in *Fundamentals of Papermaking Fibres*, ed. Balam, F., British Paper and Board Makers Assoc., Surrey (1958), 55–58.
5. Preston, R.D., *Int. Rev. Cytology*, **8** (1959), 33–60.
6. Frey-Wyssling, A. and Mühlethaler, K., *Makromol. Chem.*, **62** (1963), 25–30.
7. Jeffries, R., *J. Appl. Pol. Sci.*, **8** (1964), 1213–1220.
8. Jeffries, R., Jones, D.M., Roberts, J.G., Selby, K., Simmens, S.C. and Warwicker, J.O., *Cell. Chem. Technol.*, **3** (1969), 255–274.
9. Mühlethaler, K., *J. Pol. Sci., C*, **28** (1969), 305–316.
10. Colvin, J.R., *High Polym.*, **5** (1971), 695–718.
11. Rowland, S.P. and Roberts, E.J., *J. Pol. Sci., A-1*, **10** (1972), 2447–2461.
12. Frey-Wyssling, A., *Science*, **119** (1954), 80–82.
13. Krässig, H., *Tappi*, **61** (1978), 93–96.
14. Manley, R. St. J., *J. Pol. Sci., A-1*, **1** (1963), 1875–1892.
15. Lindenmeyer, P.H., *S.P.E. Trans.*, **4** (1964), 157–164.
16. Lindenmeyer, P.H., *Science*, **147** (1965), 1256–1262.
17. Mark, R.E., Kaloni, P.N., Tang, R.C. and Gillis, P.P., *Text. Res. J.*, **39** (1969), 203–212.
18. Muggli, R., Elias, H.G. and Mühlethaler, K., *Makromol. Chem.*, **121** (1969), 290–294.
19. Bourret, A., Chanzy, H. and Lazaro, R., *Biopolymers*, **11** (1972), 893–898.
20. Mann, J. and Marrinan, H.J., *Trans. Far. Soc.*, **52** (1956), 481–487.
21. Howsmon, J.A. and Sisson, W.A. in *Cellulose and Cellulose Derivatives*, eds. Ott, E., Spurlin, H.M. and Grafflin, M.W., 2nd edn., Interscience, New York, vol. 1 (1954), 231–316.

22. Frilette, V.J., Hanle, J. and Mark, H. *J. Amer. Chem. Soc.*, **70** (1948), 1107–1113.
23. Howsmon, J.A. *Text Res. J.*, **19** (1949), 152–162.
24. Tripp, V.W., in *Cellulose and Cellulose Derivatives*, eds Bikales, M. and Segal, L., Wiley-Interscience, New York, vol. 5 (1971), 305–323.
25. Mann, J., in *Methods in Carbohydrate Chemistry*, ed. Whistler, R.L., Academic Press, New York, vol. 3 (1963), 114–119.
26. Hermans, P.H. and Weidinger, A., *J. Appl. Phys.*, **19** (1948), 491–506.
27. Hermans, P.H. and Weidinger, A., *J. Pol. Sci.*, **4** (1949), 135–144.
28. Hermans, P.H. and Weidinger, A., *Text. Res. J.*, **31** (1961), 558–571.
29. Viswanathan, A. and Venkatakrishnan, V., *J. Appl. Pol. Sci.*, **13** (1969), 785–795.
30. Shenouda, S.G. and Viswanathan, A., *J. Appl. Pol. Sci.*, **15** (1971), 2259–2275.
31. Wakelin, J.H., Virgin, H.S. and Crystal, E., *J. Appl. Phys.*, **30** (1959), 1654–1662.
32. Hess, K., Krässig, H. and Grundermann, J., *Zeit. Physik. Chem.*, **49** (1941), 64–82.
33. Patil, N.B., Dweltz, N.E. and Radhakrishnan, J., *Text. Res. J.*, **32** (1962), 460–471.
34. Clark, G.L. and Terford, H.C., *Anal. Chem.*, **27** (1955), 888–895.
35. Segal, L., Creely, J.J., Martin, A.E. and Conrad, C.M., *Text. Res. J.*, **29** (1959), 786–794.
36. Ant-Wuorinen, O. and Visapää, A., *Norelco Reptr.*, **9** (1962), 47–52.
37. Ellefsen, O., Kringstad, K. and Tonnesen, B.A., *Norsk Skogind.*, **18** (1964), 419–429.
38. Hermans, P.H., *J. Pol. Sci.*, **1** (1946), 162–173.
39. Orr, R.S., Weiss, L.C., Moore, H.B. and Grant, J.N., *Text. Res. J.*, **25** (1955), 592–600.
40. Hermans, P.H., *Physics and Chemistry of Cellulose Fibres*, Elsevier, Amsterdam (1949).
41. Rouselle, M.A., Nelson, M.L., Hassenboehler, C.B. and Legendre, D.C., *Text. Res. J.*, **46** (1976), 304–310.
42. Nelson, M.L. and O'Connor, R.T., *J. Appl. Pol. Sci.*, **8** (1964), 1311–1324.
43. Nelson, M.L. and O'Connor, R.T., *J. Appl. Pol. Sci.*, **8** (1964), 1325–1341.
44. Pandey, S.N. and Iyengar, R.L.N., *Text. Res. J.*, **38** (1968), 675–677.
45. Nickerson, R.F., *Ind. Eng. Chem.*, **33** (1941), 1022–1027.
46. Nickerson, R.F., *Adv. Carb. Chem.*, **5** (1950), 103–126.
47. Conrad, C.C. and Scroggie, A.G., *Ind. Eng. Chem.*, **37** (1945), 592–598.
48. Phillip, H.J., Nelson, M.L. and Ziifle, H.M., *Text. Res. J.*, **17** (1947), 585–596.
49. Bobbitt, J.M., *Adv. Carb. Chem.*, **11** (1956), 1–41.
50. Guthrie, R.D. *Adv. Carb. Chem.*, **16** (1961), 105–158.
51. Goldfinger, G., Mark, H. and Siggia, S., *Ind. Eng. Chem.*, **35** (1943), 1083–1086.
52. Timell, T., *Studies on Cellulose Reactions*, Stockholm Royal Inst. Technol. (1950), 30–43.
53. Cousins, E.R., Bullock, A.L., Mack, C.H. and Rowland, S.P., *Text. Res. J.*, **34** (1964), 953–959.
54. Tarkow, H., *Tappi*, **33** (1950), 595–599.
55. Nickerson, R.F., *Text. Res. J.*, **21** (1951), 195–202.
56. Marchessault, R.H. and Howsmon, J.A., *Text. Res. J.*, **27** (1957), 30–41.
57. Rowland, S.P. and Pittman, P.F., *Text. Res. J.*, **35** (1965), 421–428.
58. Guthrie, J.D. and Heinzelman, D.C. *Text. Res. J.*, **40** (1970), 1133.
59. Guthrie, J.D. and Heinzelman, D.C. *Text. Res. J.*, **44** (1974), 981–985.
60. Dechant, J., *Faserforsch. Text. Tech.*, **18** (1967), 263–265.
61. Rouselle, M.A. and Nelson, M.L., *Text. Res. J.*, **41** (1971), 599–604.
62. Rowen, J.W. and Plyler, E.K., *J. Res. Nat. Bur. Stds.*, **44** (1950), 313–320.
63. Mann, J. and Marrinan, H.J., *Trans. Far. Soc.*, **52** (1956), 487–492.
64. Marrinan, H.J. and Mann, J., *Trans. Far. Soc.*, **52** (1956), 492–497.
65. Knight, J.A., Lamar Hicks, H. and Stephens, K.W., *Text. Res. J.*, **39** (1969), 324–328.
66. Almin, K.E., *Svensk. Papp-Tidn.*, **55** (1952), 767–770.
67. Marrinan, H.J. and Mann, J., *J. Appl. Chem.*, **4** (1954), 204–211.
68. Gladding, E.K. and Purves, C.B., *Tech. Assoc. Papers*, **26** (1943), 119.
69. Glegg, R.E., *Text. Res. J.*, **21** (1951), 143–148.
70. Harris, C.A. and Purves, C.B. *Paper Trade J.*, **110** (1940), 29–33.
71. Assaf, A.G., Haas, R.H. and Purves, C.B., *J. Amer. Chem. Soc.*, **66** (1944), 59–66.
72. Minhas, P.S. and Robertson, A.A., *Text. Res. J.*, **37** (1967), 400–408.
73. Joarder, G.K. and Rowland, S.P., *Text. Res. J.*, **37** (1967), 1083–1084.
74. Schwertassek, K., *Faserforsch. Textiltech.*, **3** (1952), 251–257.
75. Jeffries, R., Roberts, J.G. and Robinson, R.N., *Text. Res. J.* **38** (1968), 234–244.
76. Nelson, M.L., Rouselle, M.A., Cangemi, S.J. and Trouard, P., *Text. Res. J.*, **40** (1970), 872–880.

77. Doppert, H.L., *J. Pol. Sci., A-2,* **5** (1967), 263–270.
78. Wadsworth, L.C. and Cuculo, J.A., in *Modified Cellulosic,* eds Rowell, R.M. and Young, R.A., Academic Press, New York (1978), 117–146.
79. Heritage, K.J., Mann, J. and Roldan-Gonzales, L., *Text. Res. J.,* **24** (1954), 822–827.
80. Lewin, M. and Ben-Bassat, A., *1st Int. Symp. Cotton Text. Res.,* Inst. Text. de France (1969), 535.
81. Lewin, M., Guttman, H. and Saar, N., *Appl. Pol. Symp.,* **28** (1976), 791–808.
82. Lewin, M., in *Cellulose and its Derivatives,* eds Kennedy, J.F., Phillips, G.O., Wedlock, D.J. and Williams, P.A., Ellis Horwood Ltd, Chichester (1985), 27–35.
83. Hermans, P.H., *Contributions to the Physics of Cellulose Fibres,* Elsevier, Amsterdam (1946).
84. Urquhart, A.R. and Williams, A.M., *J. Text. Inst.,* **16** (1925), T138–T166.
85. Gibbons, G.C., *J. Text. Inst.,* **44** (1953), T201–T208.
86. Jeffries, R., *J. Appl. Pol. Sci.,* **12** (1968), 425–435.
87. Wadsworth, L.C. and Cuculo, J.A., *Text. Res. J.,* **49** (1979), 424–427.
88. Hunt, C.M., Blaine, R.L. and Rowen, J.W., *Text. Res. J.,* **20** (1950), 43–50.
89. Forziati, F.H., Brownell, R.M. and Hunt, C.M., *J. Res. Nat. Bur. Stds.,* **50** (1953), 139–145.
90. Merchant, M.V., *Tappi,* **40** (1957), 771–781.
91. Sommers, R.A., *Tappi,* **46** (1963), 562–569.
92. Valentine, L., *Chem. and Ind.,* (1956), 1279–1280.
93. Green, J.W., in *Methods in Carbohydrate Chemistry,* ed. Whistler, R.L., Academic Press, New York, vol. 3 (1963), 95–103.
94. Howell, R. and Jackson, A., *J. Chem. Soc.,* (1937) 979–982.
95. Klug, E.D., *J. Pol. Sci., C,* (1971) 491–508.
96. Mann, J. and Sharples, A., in *Methods in Carbohydrate Chemistry,* ed. Whistler, R.L., Academic Press, New York, vol. 3 (1963), 108–113.

3 Electrokinetics of the papermaking industry

T. LINDSTRÖM

3.1 Introduction

This chapter deals with the colloidal properties of some papermaking raw materials and particularly with their electrokinetic behaviour and its relation to the surface charge characteristics of these materials. Papermaking colloidal systems include cellulose fines and fibres, mineral fillers and pigments, natural and synthetic additives as well as dissolved or colloidal material originating from the raw materials used in papermaking. A thorough understanding of the electrokinetics of papermaking materials is essential in order to be able to control the retention and dewatering agents added to the paper machine wet end and to control the adsorption and deposition of a number of functional chemicals such as internal sizing agents, dry- and wet-strength resins, etc. This chapter briefly summarises the origin and some structural features of the electrical double layer.

The fundamentals of electrophoresis, electro-osmosis and streaming potential measurements, and of colloid titrations are also included so that the reader is acquainted with some of the more popular tools used in practical papermaking systems for surface potential and charge determinations.

3.2 The electrical double layer [1–4]

When a charged surface is immersed in a solution containing ions, an electrical double layer is formed. This is shown in Figure 3.1 for the situation close to a negatively charged surface. The charges at the particle surface may originate from one or several of the following sources:

- Dissociation of ionic groups on the particle surface (e.g. COOH-groups on cellulose).
- Adsorption of ions (e.g. Ca^{2+} ions on $CaCO_3$).
- Isomorphous substitution in the crystal lattice (e.g. in clays).

The negatively charged surface in Figure 3.1 attracts oppositely charged ions (counter-ions) and repels ions with the same charge (co-ions). If it were possible to separate the two phases, they would carry equal and opposite

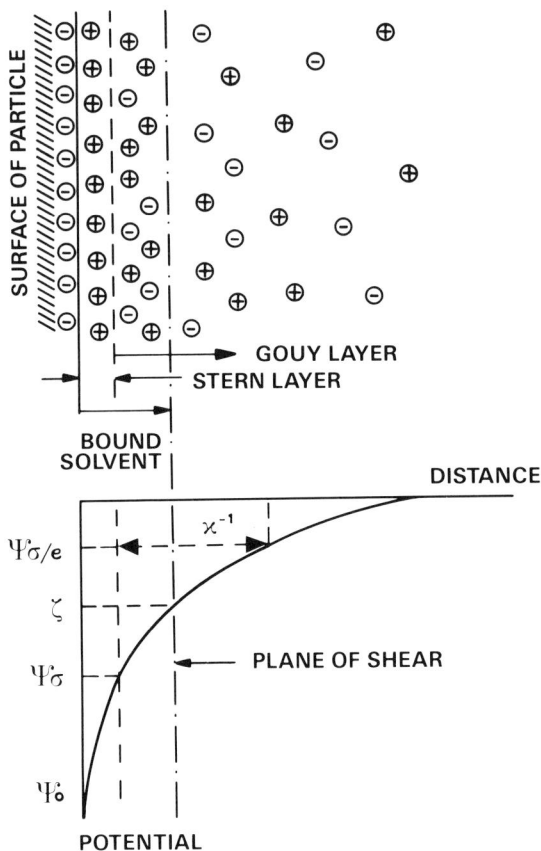

Figure 3.1   The electrostatic double layer.

charges. The two charged portions of the interfacial region are called the electrical double layer and act in principle as a capacitor.

Outward from the particle surface, the potential difference (ψ) from the bulk medium declines from ψ_0 to zero. The zeta potential (ζ) of the surface, determined from electrokinetic measurements, is not equal to ψ_0 but is, in absolute values, lower. The probability of finding an ion at a given position in the electrical double layer is determined by the Boltzmann distribution law:

$$n_i = n_{i0} \exp(-z_i e \psi / kT) \tag{3.1}$$

where n_i is the local and n_{i0} the average ion concentration, e the electron charge, ψ the potential, k the Boltzmann constant and T the absolute temperature. The charge density ρ^* is related to the ion concentration by the expression:

$$\rho^* = \sum_i z_i e n_i = \sum_i z_i e n_{i0} \exp(-z_i e \psi / kT) \tag{3.2}$$

The charge density is related to the potential through Poissons's law, which for a flat surface may be written:

$$d^2 \psi / dx^2 = -\rho^* / \varepsilon \tag{3.3}$$

where ε is the dielectric constant of the medium in the electrical double layer. Equations (3.1) and (3.3) can be combined giving the Poisson–Boltzmann equation:

$$d^2 \psi / dx^2 = e/\varepsilon \sum_i z_i e n_{i0} \exp(-z_i e \psi / kT) \tag{3.4}$$

Equation (3.4) does not have an explicit algebraic solution, but may be solved for certain limiting cases.

Introduction of the Debye–Hückel approximation, $z_i e \psi \ll kT$, i.e $\psi \ll 25.7 \, \text{mV}$ at 25°C for monovalent electrolytes, and expansion of the exponentials in eqn (3.4), yields the expression:

$$d^2 \psi / dx^2 = \kappa^2 \psi \tag{3.5}$$

where $1/\kappa$, the inverse of the so-called Debye-parameter, κ, is given by:

$$1/\kappa = (\varepsilon kT / 2e^2)^{1/2} \, 1/\sqrt{I} \tag{3.6}$$

and I is the ionic strength given by:

$$I = \tfrac{1}{2} \sum_i z_i^2 n_{i0} \tag{3.7}$$

Equation (3.5) has the solution:

$$\psi = \psi_0 \exp(-\kappa x) \tag{3.8}$$

This solution of the Poisson–Boltzmann equation is strictly applicable only in the case of low potentials, but its simplicity nevertheless offers the reader a

Table 3.1 Thickness of the electrical double layer for monovalent electrolytes in water at 25°C

C (mol/l)	$1/\kappa$ (nm)
10^{-5}	96
10^{-4}	30
10^{-3}	9.6
10^{-2}	3.0
10^{-1}	0.96

simple understanding of the features of the electrical double layer. One important aspect of this solution is the occurrence of the term l/κ in eqn (3.6) This term indicates the order of magnitude of the spatial extension of the ionic atmosphere at a particle interface. The thickness of the electrical double layer for a monovalent electrolyte at different electrolyte concentrations is listed in Table 3.1.

3.3   Origin of the charge on papermaking fines, fibres, mineral pigments and fillers

Purified cellulose is negatively charged throughout the whole pH-regime. Without any cationic additives, the electrophoretic mobility (or zeta potential) is approximately zero at pH-values $\leqslant 2.7$. Figure 3.2 shows the electrophoretic mobility of microcrystalline cellulose at different pH-values with and without different amounts of added aluminium sulphate. The negative charge on purified cellulosic surfaces originates from carboxyl groups which, depending on pH and on the type of counter-ion, dissociate to various extents at different pH-values [5].

The ionisable groups on cellulosic fibres may be carboxyl groups, sulphonic acid groups, catecholic groups, phenolic groups or hydroxylic groups. Under normal papermaking conditions, the carboxyl and sulphonic acid groups are the major contributors to the fibre charge.

Most of the carboxyl groups originate either from non-cellulosic components in the wood itself or are created during the pulping and bleaching operations. Sulphonic acid groups are introduced with the sulphite treatment during Chemi(Thermo)Mechanical Pulping (CTMP) or chemical pulping.

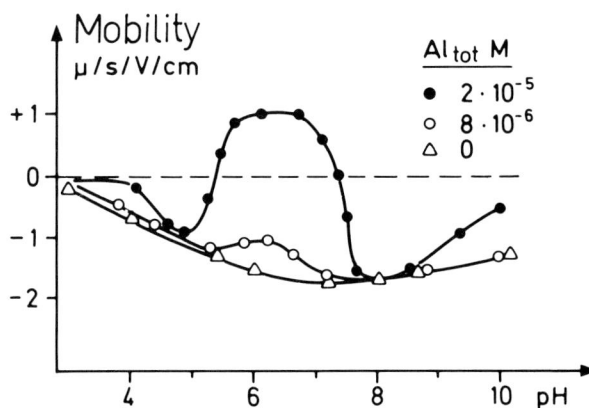

Figure 3.2 Electrophoretic mobility of microcrystalline cellulose at different pH-values with and without different aluminium substrate additions (Martin-Löf *et al.*, courtesy of the Swedish Pulp and Paper Research Institute (STFI)).

In native wood and in (Thermo)Mechanical Pulps (TMP), the carboxyl groups stem from uronic acid residues. Most of these groups are present as 4-O-methyl-α-D-glucopyranosyluronic acid bound to the xylan, although the xylan chain also contains minor amounts of α-D-galactopyranosyluronic acid groups. Some carboxyl groups are also present in the pectic substances in the fibre cell wall. The number of methylglucuronic acid groups varies in different woods but is usually of the order of 5–15 meq/100 g.

During alkaline pulping, peeling reactions are stopped by the formation of metasaccharinic acid carboxyl groups, but they constitute relatively few carboxyl groups compared with methyl-glucuronic acid groups in the xylans. During oxidative bleaching, carboxyl groups may be introduced onto the polysaccharide if the bleaching is performed under unfavourable conditions. This is, however, seldom the case during the technical bleaching of paper grade pulps.

For minerals and pigments, isomorphous substitution of ions in the crystal lattice is important (e.g. for clays), whereas for sparsely soluble materials, e.g. $CaCO_3$, the adsorption of ions at the particle interface is important.

It is important to realise that most fillers and pigments are amphoteric in nature, that is they simultaneously contain both cation and anion exchange sites. The relative ratio of cation to anion exchange sites depends on the pH and the chemical environment.

For pure clays, the cation exchange capacity originates from isomorphous replacement of Al for Si in the kaolinite structure, producing an excess negative charge [6]. Secondly, kaolinite particles are coated with a gel-like material, rich in silica. Their silanol groups dissociate to various extents depending on the chemical environment. The anion exchange capacity is attributed to anion adsorption at the particle edges. Under acidic conditions the edges are positively charged through protonation of hydroxyl groups attached to Al atoms at the platelet edges.

The anion (Cl^-) and cation (Na^+) adsorption on kaolin thus changes with pH in a manner shown in Figure 3.3. For oxide surfaces, hydrogen and hydroxide ions are important potential determining ions, and the following equilibria may for example be written for α-Al_2O_3.

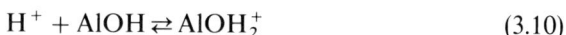

$$AlOH \rightleftarrows AlO^- + H^+ \qquad (3.9)$$

$$H^+ + AlOH \rightleftarrows AlOH_2^+ \qquad (3.10)$$

Protonation in the acidic region thus leads to cationic charges and deprotonisation in the alkaline region leads to negative charges. In Table 3.2 the isoelectric point is given for some minerals used in papermaking. It is important to realise, however, that IEP-values determined for pure minerals may be different from those of commercially supplied materials, because of either surface modification or contamination (e.g. adsorption of phosphates, silicates at the interface).

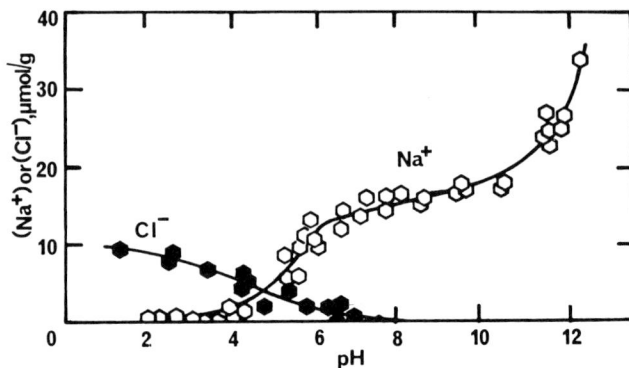

Figure 3.3   Change in Na$^+$ ion and Cl$^-$ ion adsorption on kaolin with pH [7].

Table 3.2   The isoelectric point of some mineral surfaces [5]

Mineral	IEP/pH
TiO$_2$	
anatase	6.0 ± 0.3
rutile	6.7 ± 0.1
α-alumina	9.3
SiO$_2$	2.0
Kaolinite	2.0
CaCO$_3$-calcite	8.3 ± 0.1

3.4   The classical coagulation theory, DLVO theory [1–4]

The classical coagulation theory, the DLVO theory (Derjaguin–Landau–Verwey–Overbeek), takes into account the fact that not only electrostatic repulsive forces but also London–Van der Waals attractive forces exist between dispersed particles. In this theory the total energy of interaction between two particles is a superposition of the electrostatic and the attractive force:

$$V(d) = V_E(d) + V_A(d) \tag{3.11}$$

The theory essentially predicts that, if these forces are added, the resultant has a maximum – i.e. there is a potential barrier which a particle must surmount in order to collide and unite with a second particle. At large distances between two particles, the electrostatic repulsion dominates, whereas at smaller distances the attraction dominates. The height of this barrier depends upon electrolyte concentration and counter-ion valency in a manner capable of explaining the existence of a critical coagulation concentration of an electrolyte and its counter-ion dependency. In its simplest quantitative formulation, the interaction energy between flat plates has the following form

$$V(d) = 64nkT/\kappa \, \Upsilon^2 \exp(-\kappa d) - A/12\pi d^2 \qquad (3.12)$$

$$\Upsilon = \frac{\exp(ze\psi/2kT) - 1}{\exp(ze\psi/2kT) + 1} \exp(\kappa d) \qquad (3.13)$$

where the first term is the electrostatic repulsive term and the second term the London–Van der Waals attraction term. A is the Hamaker constant of the material, and d the interparticle distance.

The attractive forces can be grouped into the following types:

- Permanent dipole–induced dipole interaction (Debye)

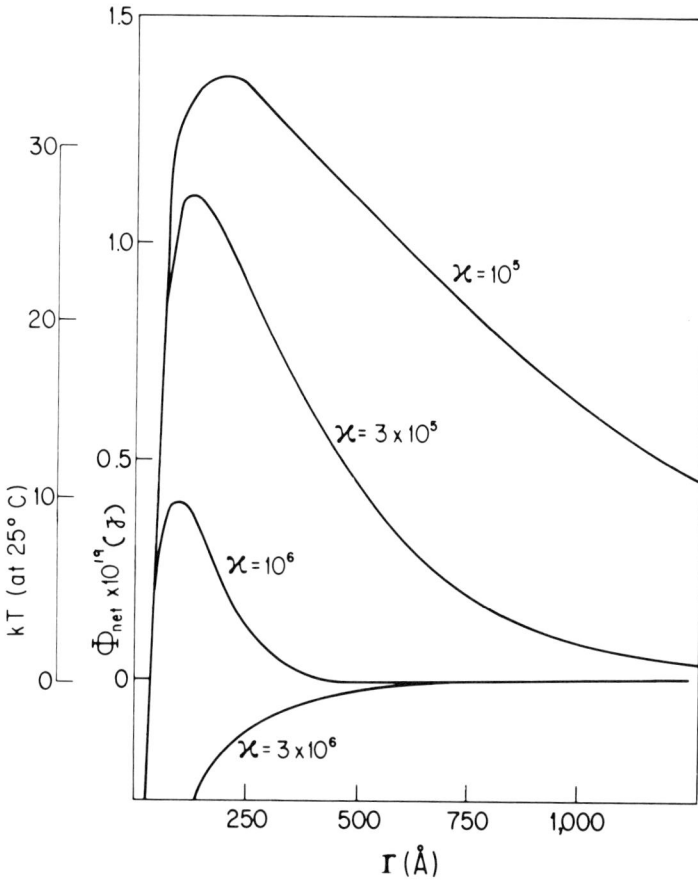

Figure 3.4 Energy of interaction between spherical particles ($r = 0.1 \, \mu$m) at different κ-values and constant A_{121} (10^{-19} J) and ψ_0 (25.7 mV) [1].

- Permanent dipole–permanent dipole interaction (Keesom)
- Induced dipole–induced dipole interaction (London)

For polar materials, Keesom forces dominate, whereas London forces dominate in the interaction between non-polar materials.

For individual atoms, these attractive forces decay rapidly with increasing interatomic distance ($\sim r^{-7}$) but when they are summed for the interaction between particles in the colloidal size range they are of significant wide range ($\sim r^{-2}$) and strength. It is convenient to designate a cluster of constants as the Hamaker constant, A.

Using equation (3.12) and appropriate constants, the energy of interaction between particles may be calculated. A typical interaction curve for two spherical particles of the same radius (1000 Å) is given in Figure 3.4. Particles having a thermal energy ($k_B T$) exceeding the barrier will be intercepted (coagulation during a collision). The DLVO theory essentially predicts:

- The higher the potential at the surface of a particle, the greater will the repulsion be.
- The larger the Hamaker constant, the greater is the attraction between molecules and, by extension, between macroscopic bodies.
- The lower the electrolyte concentration, the greater is the distance from the surface before the repulsion drops significantly.

It has long been known that the addition of an electrolyte causes coagulation of lyophobic colloids, and this is also the case for finely dispersed cellulose sols. The higher the valency of the counter-ions, the greater is their ability to compress the electrostatic double layer. This is the qualitative statement of the Schulze–Hardy rule. Generally, the DLVO theory predicts that the relationship between the C.C.C. (Critical Coagulation Concentration) and the valency of the counter-ion should be C.C.C. $\sim Z^{-6}$.

3.5 Electrophoresis and electrokinetic phenomena – calculation of the zeta potential [1–4]

The word 'electrokinetic' implies the combined effect of motion and electrical phenomena. Electrokinetic phenomena involve processes where there is a relative velocity between two parts of the electrical double layer. This may arise from the movement of charged particles in an electric field (electrophoresis) or from the movement of the solution phase relative to a stationary wall (electro-osmosis or streaming potential). In each case, electrokinetic measurements involve a quantity known as the zeta potential (ζ).

The zeta potential is the potential in the slip plane or surface of shear between the two phases (see Figure 3.1). The exact location of this slip plane is more or less ill defined, although it is often assumed that $\zeta \approx \psi \sigma$, because the bound solvent layer in the time frame of an electrophoresis experiment is probably very thin, consisting at most of a few molecular layers of bound solvent.

The task is to convert electrophoretic mobilities, the experimental quantity in an electrophoresis experiment, electro-osmosis or the streaming potential, to the zeta potential of a surface. A working relationship can easily be derived for certain limiting cases.

Consider the case where the thickness of the double layer is thin compared with the radius of curvature (i.e. flat surface or $R\kappa \gg 1$). Let us also consider the force balance for an infinitesimal element in the electrostatic double layer. If the electrostatic force ($= \bar{E}\rho^*$, $\bar{E} =$ field strength, ρ^* charge density) is equated with the net viscous drag force ($\eta \nabla^2 \vartheta$) of the element, the following equation may be written (Stokes equation with $\Delta \rho = 0$):

$$\eta \, d^2\vartheta/dx^2 = -\rho^* E \qquad (3.14)$$

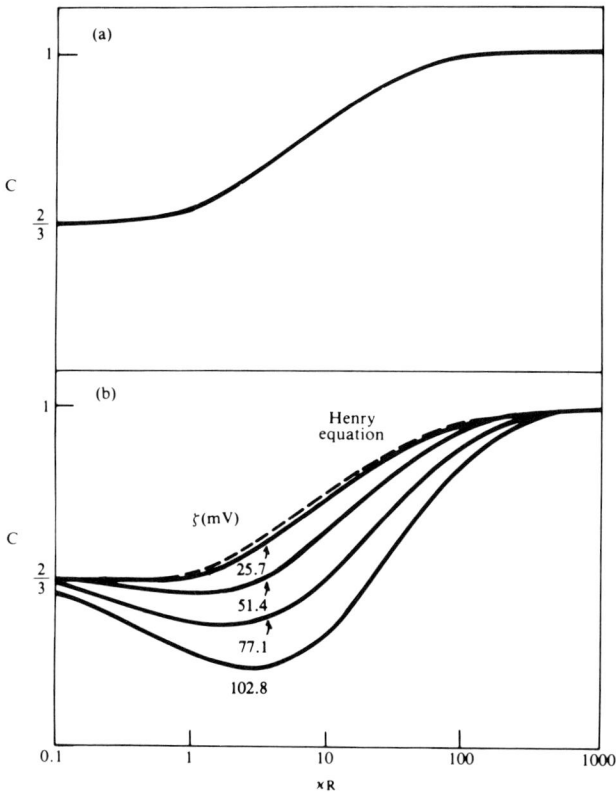

Figure 3.5   Variation of the constant f with κR and the zeta potential [1].

Substituting ρ^* in Poisson's eqn (3.3) gives:

$$\eta \, d^2 \vartheta/dx^2 = E \varepsilon \, d^2 \psi/dx^2 \qquad (3.15)$$

Equation (3.15) may easily be integrated with the following boundary conditions:

$$\psi = \zeta; \; \vartheta = 0$$

$$\psi = 0; \; \vartheta = \vartheta e$$

$$d\psi/dx = 0; \; x \to \infty$$

giving the following working relationship:

$$\vartheta/E = Ue = \varepsilon \zeta/\eta \qquad (3.16)$$

where Ue is the electrophoretic mobility. Equation (3.16) is called the Helmholtz–Smoulochowski (HS) equation and it has been shown that this result is valid for $\kappa R > 100$.

For intermediate cases the HS equation must be somewhat modified. The HS equation can be written in the form:

$$Ue = f \varepsilon \zeta/\eta \qquad (3.17)$$

where $f = 1$ for $\kappa R > 100$ and $f = 2/3$ for $\kappa R < 0.1$. For intermediate values of κR, f varies with κR and the zeta potential according to Figure 3.5. There is also an effect of particle shape. Provided the electrostatic double layer is thin compared with the particle size, the expression for the relationship between electrophoretic mobilities and zeta potentials (eqn 3.17) remains the same with $f = 1$. For practical purposes it is not necessary to convert electrophoretic mobilities into zeta potentials as long as particle electrophoresis measurements are made. As soon as these measurements are compared with electro-osmosis or streaming potential measurements the relationships between the experimental quantities and the zeta potential are useful.

3.6 Some experimental methods of determining the zeta potential

The zeta potential may be determined by one or more of the four principal electrokinetic methods; electro-osmosis, electrophoresis, streaming potential and sedimentation potential (Dorn-effect). The four methods are related to each other as shown in Table 3.3. The principal experimental set-ups are illustrated in Figure 3.6.

3.6.1 *Microelectrophoresis*

Electrophoresis is by far the most common procedure for the determination of the zeta potential. In microelectrophoresis, the movement of individual colloidal particles under the influence of a known electrical field is followed

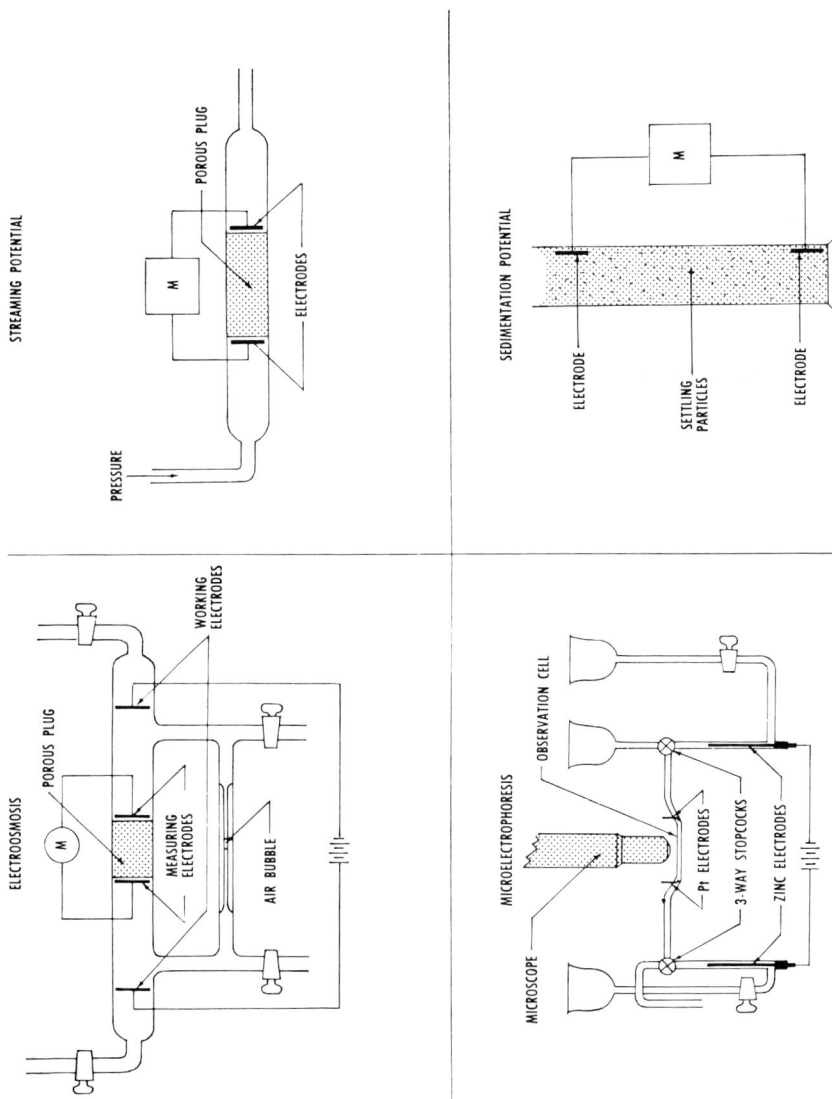

Figure 3.6 Illustration of the four principal ways of zeta potential measurements [8].

Table 3.3 Methods for zeta-potential determinations

	Application of an electrical field and measurement of liquid or particle movement	Liquid or particle movement and measurement of an electrical potential
Particle movement	Electrophoresis	Sedimentation potential
Movement of a liquid along a phase boundary	Electro-osmosis	Streaming potential

directly in an ultramicroscope assembly (or dark-field microscope). If the field strength is known in the area of observation, the electrophoretic mobility (μm/s/V/cm) is easily calculated from the observed particle velocity. The zeta potential is then calculated using eqn (3.16). The particle concentration should be as low as possible to avoid interparticle interference. For measurements on papermaking suspensions, the fibre material must first be screened off.

The cells used may have a circular or flat cross-section, although the latter is the preferred cell configuration in order to minimise optical corrections. The measurements take place in a closed capillary and both the particles and the liquid are affected by the imposed electrical field. As the cell surface (usually glass) is charged (usually negatively), the counter-ions in the electrostatic double layer at the cell wall interface move towards the oppositely charged electrode under the influence of the electric field. As there can be no net

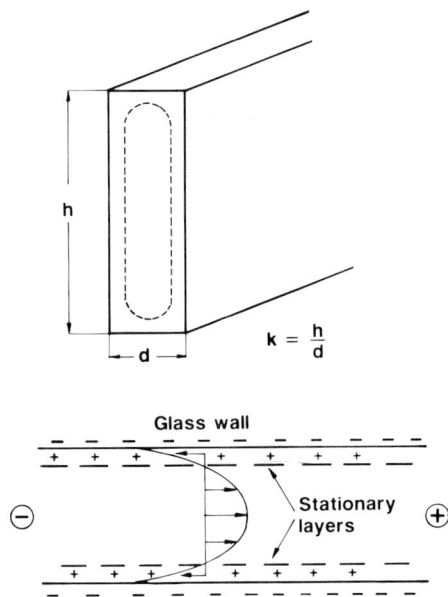

Figure 3.7 Positioning of the stationary layers in the flat type of microelectrophoresis cell.

movement of liquid in the cell there is backflow in the middle of the cell (Figure 3.7).

From the Poiseuille velocity flow profile, the stationary levels in the cell can be calculated. With flat cells, the stationary layers are located at a distance X_0 from the internal cell surface as given by the following equation:

$$X_0/d = 0.5 \pm [1/12 + (2/\pi)^5 1/k]^{1/2} \qquad (3.18)$$

where d is the cell thickness and k the ratio of breadth to depth of the cell. It is preferable to achieve a large k by increasing the breadth rather than the depth, because a very shallow cell produces a very sharp velocity profile with a consequent increase in error due to the finite depth of field of the microscope objective used for the mobility measurement. It is also advisable to check that the particle mobilities are the same at the two stationary layers.

Experimentally it is important to avoid convection currents caused by temperature gradients in the fluid. This is most easily checked by ensuring that the particles stand still when the electrode voltage has been switched off. Platinised electrodes are satisfactory and convenient for work at low electrolyte concentrations. The platinum black layer is deposited on the electrode by electrolysis in a chloroplatinic acid solution containing a little lead acetate. It is advisable to reverse the electrode polarity frequently in order to avoid hydrogen gas evolution at the anode. At high electrolyte concentrations it is necessary to use reversible electrodes of the Ag/AgCl/KCl $(0.1–1.0 \text{ M})$ or $Cu/CuSO_4$ $(0.1–1.0 \text{ M})$ type.

3.6.2   Electro-osmosis

In electrophoresis, the dispersed phase moves and the continuous phase is stationary when an external electrical field is applied. It is obvious that the required relative motion between a surface and its double layer can also be brought about by causing the electrolyte solution to flow past a stationary charged cell wall. The complements of electrophoresis are electro-osmosis and streaming potential. In electro-osmosis an applied external field induces the solution flow, which is measured. In streaming potential measurements, the solution is made to flow by applying a pressure and a potential is induced as a result. Cause and effect are thus interchanged in electro-osmosis and streaming potential.

The electro-osmosis apparatus shown in Figure 3.6 consists of a porous plug (e.g. a plug made of cellulosic fibres), two working electrodes and two measuring electrodes, to record the electrical field over the plug. The fluid motion is measured by means of a measuring capillary containing an air bubble to indicate fluid displacement. When an electric field is applied over the plug, the double layer ions begin to migrate and a stationary state is soon reached where the electrical and viscous forces balance each other. The

tangential displacement of the fluid relative to the wall defines a surface of
shear at which the potential equals the zeta potential.

Equation (3.16) relating the electrophoretic mobility to the zeta potential
can also be used in electro-osmosis. The volume of liquid (V) displaced per unit
time is given by multiplying both sides of eqn (3.16) with the cross-sectional
area of the plug. This equation for the working relationship for electro-osmosis
may then be written:

$$V/E = \varepsilon \zeta A/\eta \qquad (3.19)$$

Using Ohm's law relating voltage to current ($E = I/A\lambda_0$ where λ_0 is the electrical conductivity of the fluid), eqn (3.19) may also be written in the form:

$$V/I = \varepsilon \zeta/\eta \lambda_0 \qquad (3.20)$$

In electro-osmosis, the electric field should be applied through a reversible
electrode system because it is necessary to pass current in one direction for
quite a lengthy period to observe sufficient volume transport. The electrode
system must be capable of doing so without the gradual development of
polarisation effects which would change the field strength over the plug.

3.6.3   Streaming potential

In the experimental set-up shown in Figure 3.6, the liquid is forced through
the plug by applying a pressure difference Δp and measuring the streaming
potential, V_s. In order to derive a working equation, the Onsager principle of
reciprocity may be used. As a consequence of the thermodynamics of
irreversible processes, it may be shown that:

$$(V_s/\Delta p)_{i=0} = (V/I)_{p=0} \qquad (3.21)$$

Thus, using eqn (3.20), we can write

$$V_s/\Delta p = \varepsilon \zeta/\eta \lambda_0 \qquad (3.22)$$

The streaming potential may be measured over capillaries, thin slits or packed
beads over which a fluid is forced.

In practice it is important to avoid the development of asymmetric
potentials (a finite intercept in a plot of V_s vs Δp) or stray potentials. One way of
avoiding asymmetric potentials is to use sinusoidal measurements, that is to
use a sinusoidal pressure so that an alternating current is created. One popular
design is to force liquid back and forth through a small slit using a piston. This
type of instrument is popular for making charge reversal curves to monitor the
cationic demand of various suspensions. It must be recalled that such indirect
measurements on other surfaces (often Teflon-covered cylinder and piston) do
not necessarily reflect the zeta potential of the suspension. It appears, however,
that the point of charge reversal (when using polyelectrolytes) of the
suspension and the measuring surface are in many cases intimately related.

3.7 Polyelectrolyte titrations [9–16]

3.7.1 *General principles*

Polyelectrolyte or 'colloid titrations' are used to determine the concentration of charged polymers in solution by titration with a polymer of opposite charge. The titration is based on the metachromatic colour shift of certain dyes (e.g. orthotoluidine blue, crystal violet, acridine orange, etc.) when they form complexes with chromotropic polyelectrolytes. Figure 3.8 illustrates the metachromatic colour shift of orthotoluidine blue (OTB) when complexed with potassium polyvinylsulphate (KPVS).

The polyelectrolyte titration can be summarised as follows:

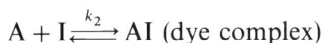

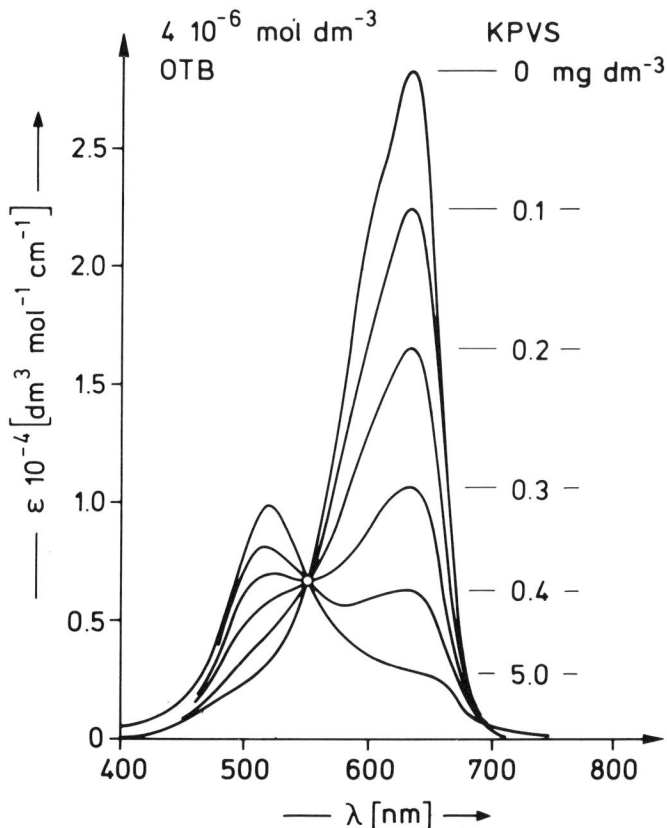

$$A + C \xrightleftharpoons{k_1} AC \text{ (polyelectrolyte complex)}$$

$$A + I \xrightleftharpoons{k_2} AI \text{ (dye complex)}$$

Figure 3.8 The metachromatic change in the absorption bands of orthotoluidine blue complexed with polyvinylsulphate KPVS [13].

where A is an anionic polyelectrolyte, C a cationic polyelectrolyte and I the indicator. The anionic polyelectrolyte is usually KPVS and the indicator is usually OTB. A necessary prerequisite for a successful titration is that the equilibrium constant k_1 is much larger than k_2, which is most often the case for polyelectrolytes.

If, for example, a cationic polyelectrolyte together with OTB is titrated with KPVS, a polyelectrolyte complex is initially formed until no free cationic polyelectrolyte is left to react with KPVS. At this point, KPVS starts to complex with OTB and a colour shift from light blue to purplish red indicates the end point of the titration.

The polyelectrolyte titration principle relies on the fact that oppositely charged polyelectrolytes form 1:1 complexes. This is generally true, provided the ionic strength is sufficiently low. It is obvious that there are physical restrictions to the extent to which polyelectrolytes can complex with each other. It is important to notice that the matching does not need to be better than the size of the electrostatic charge field surrounding the polymer backbone. We can quantify this by saying that the maximum distance between the charges in a complex must be smaller than the size of the electrostatic double layer. Polyelectrolyte titrations should thus be carried out under conditions of low ionic strength.

Polyelectrolyte titrations can naturally be performed with different polymers, both anionic and cationic. The only requirement is that the two oppositely charged polyelectrolytes shall have a sufficiently high charge density to form polyelectrolyte complexes.

The interaction between OTB and KPVS is also inhibited by high electrolyte contents. Limiting values are 5×10^{-2} M (Na$^+$), 5×10^{-3} M (Ca^{2+}) and 5×10^{-4} M (Al^{3+}).

3.7.2   *Determination of surface charge of solids by means of colloid titration*

The polyelectrolyte titration technique may also be used to determine the surface charge of solid substrates such as cellulose fibres and mineral fillers and pigments. Such titrations are called indirect polyelectrolyte titration (see Figure 3.9). In this mode a cationic or anionic polyelectrolyte is equilibrated with a solid substrate, after which the solid phase is separated from the aqueous phase and the residual polyelectrolyte in the solution phase is titrated.

A prerequisite for this indirect polyelectrolyte titration method is that the number of adsorbed charges equals the number of surface charges, i.e. the ion-exchange capacity of the solid substrate. This requires, of course, that the adsorption on the oppositely charged polyelectrolyte is governed only by electrostatic interactions. The polyelectrolyte adsorption process is generally stoichiometric only in the limit of low ionic strengths, analogous to the previously considered polyelectrolyte complexation phenomena.

Direct titration

C + A → PEC (polyelectrolyte complex).

Indirect titration

$$C(A) + \text{fibre suspension} \xrightarrow{\text{filtration}} \text{filtrate}$$

Filtrate (dil. with deion. H_2O) + A(C) → PEC

Figure 3.9 Principal schemes for different applications of the polyelectrolyte titration technique.

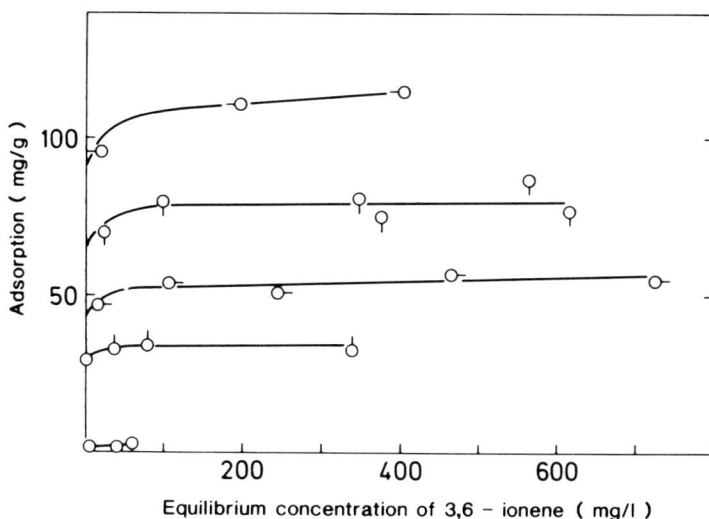

Figure 3.10 Isotherms for the adsorption of 3,6-ionene on carboxymethylated pulp fibres with different degrees of carboxymethylation [16]. Degree of substitution ○ = 0.007; ○̇ = 0.026; ○− = 0.042; ○̣ = 0.069; −○ = 0.113.

Polyelectrolyte adsorption isotherms are generally of the high affinity type and the surface charge density can be determined from the adsorption plateau if stoichiometry prevails. For good accuracy, a knowledge of the full adsorption isotherm is usually required (example, see Figure 3.10). Stoichiometry has been established in a number of cases for cellulosic fibres, provided the cationic polyelectrolyte has a sufficiently low molecular weight to be able to penetrate the swollen cell wall of the fibres and reach the charges. For cellulosic fibres with a low ion-exchange capacity, the equilibration process may take a long time and precautions have to be taken to ensure that conditions for adsorption equilibrium are fulfilled.

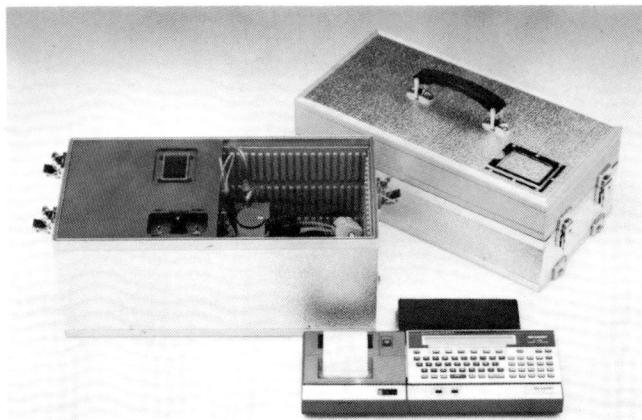

Figure 3.11 Example of commercial instrumentation for polyelectrolyte titrations using photo-electric end-point determinations (Courtesy of T. Lindström Inc., Repap Products).

3.7.3 Instrumentation of colloid titration

The end-point in a polyelectrolyte-titration experiment may be visually evaluated. It is possible to achieve a reasonable repeatability, but the reproducibility is often poor. In order to avoid human error, Horn [13] developed a measurement technique in which the difference in light transmission from two light-emitting diodes, one red and one green, is measured. The absorbance from the green diode is essentially constant during the titration as this wavelength band corresponds to the isosbestic point of the KPVS–OTB complex. The change in absorbance is thus recorded by the transmission of light from the red diode. A photograph of a commercial instrument based upon this principle is shown in Figure 3.11. The idea behind this technique is to correct for the change in turbidity originating from precipitates, etc. formed during the titration. By taking the differential between the two transmission values, it is possible to correct for the changed light transmission due to the turbidity of the sample. (The turbidity is less sensitive to the wavelength of the light within this wavelength band.)

Polyelectrolyte titrations can also be performed without chromophoric polyelectrolytes and dyes using, for instance, a streaming current detector to determine the isoelectric point during a titration.

References

1. Hiemenz, P.C., *Principles of Colloid and Surface Chemistry*, 2nd edn, Marcel Dekker, Inc., New York and Basel (1986).
2. Vold, R.D. and Vold, M.J., *Colloid and Interface Chemistry*, Addison-Wesley Publishing Company, Inc., Massachusetts (1983).

3. Hunter, R.J., *Foundations of Colloid Science*, Vol. 1, Clarendon Press, Oxford (1987).
4. Hunter, R.J., *Zeta Potential in Colloid Science*, Academic Press, London (1981).
5. Lindström, T., 'Some fundamental chemical aspects on paper forming', in *Fundamentals of Papermaking, Transactions of the Ninth Fundamental Research Symposium*, Cambridge (1989), vol. 1 ed. C.F. Baker and V.W. Punton, Mechanical Engineering Publications Ltd, London (1989), 309.
6. Jepson, W.B., *Phil. Trans R. Soc. London*, **A311** (1984), 411.
7. Ferris, A.P. and Jepson, W.B., *J. Coll. Interface Science*, **51** (1975), 245.
8. Sennet, P. and Olivier, J.P., *Ind. Eng. Chem.*, **57** (1965), 33.
9. Terayama, H., *J. Polym. Sci.*, **8** (1952), 243.
10. Michaelis, L., *Cold Spring Harbor Symp. Quant. Biol.*, **12** (1947), 131.
11. Bargeron, J.A. and Singer, M., *J. Biophys. Biochem. Cytol.*, **4** (1958), 433.
12. Gummow, B.D. and Roberts, A.F., *Makromol. Chem., Rapid Comm.*, **6** (1985), 381.
13. Horn, D., *Progr. Colloid Polymer Sci.*, **65** (1978), 251.
14. Burkhardt, C.W., McCarthy, K.J. and Parazak, D.P., *J. Polym. Sci., part C: Polymer Letters*, **25** (1987), 209.
15. Casson, D. and Rembaum, A., *Macromolecules*, **5** (1972), 75.
16. Winter, L., Wågberg, L., Ödberg, L. and Lindström, T., *J. Colloid Interface Sci.*, **111** (1986), 537.

4 Mechanisms of flocculation and stabilisation of suspensions by organic polymers

F. LAFUMA

4.1 Introduction

Disperse systems are encountered in a wide variety of fields, especially in the chemical industry where they can be used as end products (paints, printing inks, drilling muds, cosmetics, agrochemical or pharmaceutical formulations) or in one or several stages of the elaboration of materials such as ceramics, cements or coatings. Such applications require a good stability of the suspension or, if necessary, a controlled level of aggregation of the medium in order to improve its mechanical behaviour. Other important technological processes need a fast, complete or selective solid/liquid separation of disperse media: papermaking, water treatment, clarification of beverages, mineral processing, etc. A flocculation step is then essential in the case of submicronic or colloidal particles which do not settle spontaneously and cannot be filtered. Practical situations may be still more complex. For instance, in papermaking, pigments and clays are used for coating and filling. In both cases, their state of dispersion affects the quality and the requirements are different. A well dispersed system is generally necessary to apply the coating, but the final optical properties will depend on the degree of pigment aggregation. Conversely, the retention of fillers and pulp fines during sheet formation at the wet-end is better if flocculated, nevertheless, a minimum degree of dispersion is required for a good homogeneity of the sheet. Moreover, at the same time one must avoid fibre flocculation which is detrimental to paper quality.

The main characteristic feature which is common to all dispersions is the existence of a large interfacial region, the structure and properties of which drive the interparticle interactions responsible for the stability of the system. Due to their adsorption or depletion at the solid/liquid interface, polymers modify the balance between attractive and repulsive forces; consequently, their addition allows the state of the dispersion to be controlled. In fact, depending upon their chemical structure, molecular weight and concentration, they can act either as dispersants or as flocculants and they have been used for these purposes for many years. An understanding at a fundamental level of the mechanisms of their functioning is a prerequisite for the optimisation of their performances. This has become possible thanks to the existence of well-characterised model suspensions which are less complex than real, natural or industrial systems. This chapter is a general survey of the effect of adsorbed polymers on the stability of colloidal suspensions.

4.2 Behaviour of suspensions of bare particles

4.2.1 *Interparticle forces and DLVO theory*

In the absence of additives, two kinds of long range interactions are mainly operative [1]: van der Waals attractions and electrostatic repulsions to which must be added attractive effects between hydrophobic surfaces (such as pigments or dyes) in water. However, although there is no doubt about the existence of these latter forces, there is still controversy about their magnitude and range that cannot be evaluated in a quantitative way.

Van der Waals attractions are intrinsic to the particle/solvent pair. Their origin is molecular: London dispersion forces. At the macroscopic scale, the attractive energy V_A between two bodies varies as an inverse power function of their distance. For instance, for two identical spheres of radius a, separated by a distance r, the attractive energy V_A to a first approximation is:

$$V_A = -\frac{Aa}{12r}$$

A is the Hamaker constant of the particle/solvent pair which depends on the density and polarisability of the molecules constituting the particles and the solvent. A scales like an energy and displays values of a few kT for classical oxide or latex suspensions (kT is the thermal energy).

Electrostatic repulsions are induced by the occurrence of charged surface groups. These charges may have various origins such as surface group dissociation in aqueous media, electron or proton transfer in organic solvents, isomorphic substitutions in clays or specific adsorption of charged species. The existence of those charges induces a surface potential Ψ_0 and disturbs the distribution of small ions in the solution, leading to an accumulation of counterions and a depletion of co-ions in the vicinity of the surface. Repulsions occur in fact between the so-created electrostatic double-layers; their range K^{-1} is strongly dependent on the ionic strength I of the medium:

$$K^{-1} = \left(\frac{\varepsilon kT}{4N_A e^2 I}\right)^{1/2} \quad \text{with} \quad I = \frac{1}{2}\sum c_i z_i^2$$

N_A is the Avogadro number, e the electron charge, c_i and z_i the concentrations and valencies of the different ionic species in the solvent and ε its dielectric constant. So, orders of magnitude of κ^{-1} at room temperature in water, are for instance 10 nm for $I = 10^{-3}$M and 1 nm for $I = 10^{-1}$M.

A first approximation of the repulsive energy V_R between two spheres separated by a distance, r, with $\kappa a \gg 1$ and $\Psi_0 < 25$ mV is:

$$V_R = 2\pi\varepsilon a \Psi_0{}^2 \log(1 + e^{-\kappa r})$$

Therefore, unlike attractive ones, repulsive forces can always be screened by modifying external parameters such as ionic strength or even pH which often drives the dissociation of surface charges. It is important to notice that in organic media, electrostatic phenomena are often much more complex since the least polar impurity can completely change the characteristics of the interface; moreover, although they never govern in this case, electrostatic repulsions should not be systematically ruled out [2].

The basic DLVO theory was proposed about 50 years ago by Derjaguin and Landau and independently by Verwey and Overbeek; it postulates the additivity of these two kinds of interactions (V_A and V_R) together with the short-ranged Born repulsions between electronic clouds V_B. Such an approach strictly concerns lyophobic colloids at low concentration but it represents a good first approximation in many cases. Thus, the stability of a suspension mainly depends on the κ^{-1} value, i.e. on the concentration and valency of the small ions in solution. The resulting energy/distance curves (Figure 4.1) can display at low ionic strength a maximum which represents an energy barrier ΔV preventing the particles from irreversible aggregation in the primary minimum which remains the most stable thermodynamic

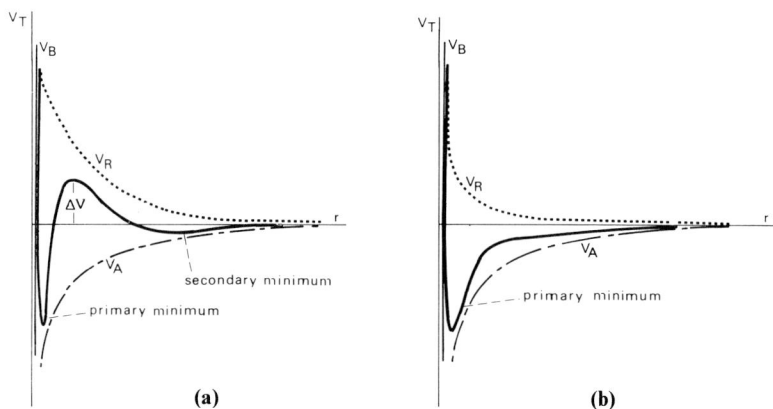

Figure 4.1 Interparticle interaction energy as a function of distance. (a) Low ionic strength; (b) high ionic strength.

state of the system. The suspension stability is then determined by the relative value of ΔV towards the thermal energy kT. Systems for which $\Delta V \gg kT$ are metastable whereas slow coagulation occurs when $\Delta V \approx kT$. The salt concentration corresponding to $\Delta V = 0$, leading to fast coagulation of the system is called critical coagulation concentration (c.c.c.).

4.2.2   *Limitations of electrostatic stabilisation/destabilisation phenomena*

One of the major drawbacks of purely electrostatic stabilisation mechanism is that systems are only metastable. Consequently, the risk of aggregation as a function of time can never be completely discarded. It also follows from the general behaviour described above that stability is seldom achieved in the absence of strong electrostatic repulsions, i.e. at high salt concentration or in organic solvent. Conversely, it is impossible to induce coagulation through electrostatic mechanisms at low ionic strength. Finally, problems of controlled aggregation remain unresolved in a general way because they would require manipulations of electrostatic repulsions with careful adjustments of I and pH.

However, at first sight, double layer compression by increasing ionic strength appears a very simple and cheap method for coagulating suspensions, especially with non-toxic salts. Calculation of the c.c.c. shows that it depends both on the surface potential and on the valency z of counterions: it is a function of z^{-6} for high Ψ_0 and z^{-2} for low Ψ_0. These results underline the importance of multivalent ions in coagulation processes because they can be efficient at very low concentration, allowing their use, e.g. for drinking water treatment and ruling out some of the above-mentioned disadvantages. In fact, only a few of them are not toxic and their maximum efficiency is often restricted to a limited concentration and pH range outside which they form complex species of lowest valencies, or precipitate as hydroxides [3]. On the other hand, phenomena can be still more complex accounting for the affinities of the small ions in solution towards the surface: indifferent ions which act by simple screening of electrostatic repulsions as described above must be distinguished from specifically adsorbed and potential determining ions, that can also induce changes in the surface charge [4].

Other drawbacks arise from the small size of the so-formed aggregates and from their poor mechanical characteristics which result in slow decantation, plugging, breakages, etc.... Therefore, in spite of the low cost of these reagents, it appears necessary to have recourse to other additives, in order to improve the stabilisation/destabilisation properties of colloidal systems. Among them, polymers display a high efficiency because their sizes are of a similar order of magnitude to the range of the interparticle surface forces.

4.3   A brief insight into polymer conformations in solution and at solid/liquid interfaces

Addition of a polymer to a colloidal suspension leads either to an increase in stability or to phase separation of the system. Both situations have been exploited for practical purposes for many years. The corresponding effects are usually irreversible and the final result depends essentially on the interfacial polymer conformation and on the level of coverage of the surface by the adsorbed macromolecules. Schematically, there are two main adsorption modes for polymers at solid/liquid interfaces: nonionic polymers in aqueous or organic solvents and polyelectrolytes oppositely charged to the surface in salt-free solution. All other situations are intermediate cases.

4.3.1   Polymer solutions

Whereas the thermodynamics of nonionic polymer solutions is now rather well understood, there is still no definite theory for highly charged polyelectrolytes. In dilute solution, flexible nonionic macromolecular chains adopt random coil conformations, the size of which depends on the solvency of the medium, i.e. on the relative polymer segment/solvent and segment/segment affinities. If the first ones dominate the second, the solvent is good, the coils are osmotically swollen and they display their maximum radius of gyration. When the solvency decreases, the chains progressively collapse and finally precipitate in bad solvents. For each polymer/solvent pair, there is a Θ temperature which corresponds to the limit of the polymer

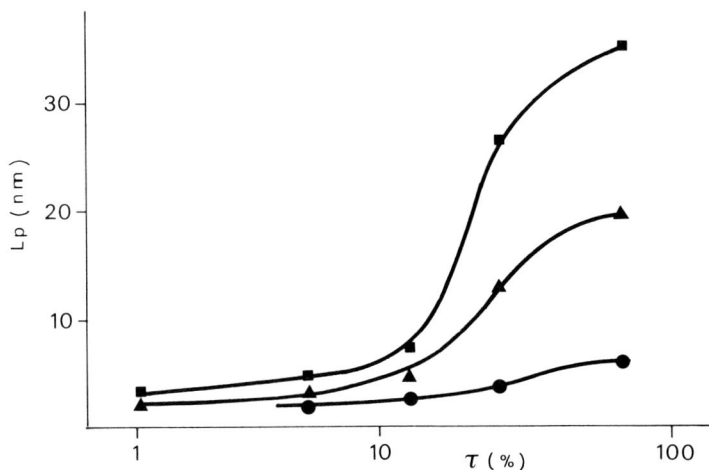

Figure 4.2   Persistence length of cationic copolymers of acrylamide as a function of the chain charge content for $I = 1 \, \text{M} \, (\bullet)$, $10^{-2} \, \text{M} \, (\blacktriangle)$ and $2 \times 10^{-3} \, \text{M} \, (\blacksquare)$. Redrawn from Ref. 5.

solubility. So, the radius of gyration of a macromolecule is a function of its molecular weight M_w and of the solvency: it varies according to a power law M_w^α, ($\alpha = 0.6$ in a good solvent, 0.5 in a Θ solvent).

The above schedule is not convenient for highly charged polyelectrolytes in salt-free solutions. Because of the intrachain electrostatic repulsions between charged units, these display more or less extended conformations according to their charge density and their chain stiffness is generally characterised by a persistence length L_p. Addition of salt screens these repulsions and allows the polymer to recover coiled conformations. So, the solution conformation of polyelectrolyte chains depends both on the percentage of charged polymer units (τ) and on the ionic strength I of the medium; this is illustrated on Figure 4.2 which displays the evolution of L_p with those two parameters for a series of random cationic copolymers of acrylamide (I) and N,N',N'' trimethylaminoethyl acrylate (II) [5].

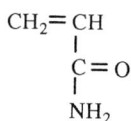

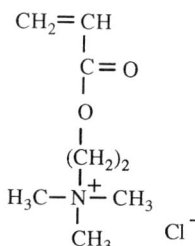

4.3.2 *Polymers at solid/liquid interfaces: general characteristics for nonionic polymers and highly charged polyelectrolytes*

In the vicinity of a solid surface, macromolecules more or less keep their solution conformation because they cannot be completely confined next to the surface. Various theories account for the phenomenological descriptions given below [4, 6, 7].

For instance in a good solvent, an uncharged flexible chain is anchored to the surface by a small fraction of its segments (trains) while most of them belong to loops and tails dangling in the solvent (Figure 4.3). At the local scale, trains adsorb through classical specific or non specific interactions (acid/base, hydrophobic or hydrogen bonding), including the displacement of solvent molecules from the surface. The adsorbed amount Γ, and the thickness δ of the adsorbed layer, depend on molecular weight and solvency. When the solvency decreases, the lateral osmotic repulsions between segments are lowered and the layers collapse while the adsorbed amount increases. Orders of magnitude are $mg \cdot m^{-2}$ for Γ and the radius of gyration for δ (e.g. approximately a few nm for $M_w = 1000$ and about 100 nm for 10^6).

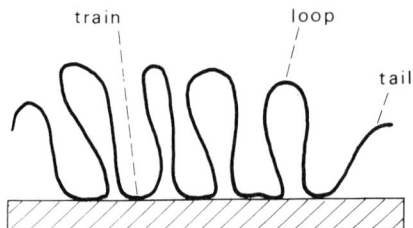

Figure 4.3 Representation of a nonionic macromolecular chain adsorbed at a solid/liquid interface.

Because of their more rigid conformation and higher affinity for the surface, oppositely, highly charged polyelectrolytes are expected to display flatter adsorbed layers than uncharged polymers and lower adsorbed amounts due to the strong lateral electrostatic repulsions between segments in the adsorbed layer. Increasing the ionic strength screens simultaneously segment/segment and segment/surface interactions: the chains become more flexible giving rise to a more loopy conformation together with changes in adsorbed amounts [8]. For high enough surface charge densities, Γ generally increases, the main effect being the screening of lateral segment/segment repulsions, but for low charged surfaces, in the absence of a possibility of a nonelectrostatic interaction, the simultaneous screening of segment/surface attractions can lead to a decrease of Γ and even to a displacement of the macromolecule from the surface at high ionic strength. The same general trends are observed when the charge density of the chain is lowered and are detailed in section 4.3.3.

Finally, the adsorption of polyelectrolytes bearing charges of the same sign as the surface groups becomes possible if a strong enough polymer/surface interaction overcomes the electrostatic repulsion for example, if the charged units are multifunctional, or if they are copolymerised with an adsorbing monomer, or if the possibility exists of complex formation between the polymer and the surface through e.g. a multivalent ion [9–10]. This adsorption obviously increases with ionic strength when unfavourable electrostatic polymer/surface repulsions are screened.

A common feature to all polymers is that the adsorption of one segment is not very energetic (a few kT or less according to the kind of interaction [11]). For a whole macromolecular chain, the number of train segments can be very large, especially for high degrees of polymerisation, and so is the adsorption energy. That is why polymer adsorption is generally irreversible (whilst adsorption is reversible at the level of the monomeric unit). The consequences are as follows:

(i) the permanent character of stabilisation/destabilisation phenomena is induced by polymeric additives;

(ii) the replacement of a macromolecule adsorbed onto a solid surface can be achieved segment by segment if sufficient concentration of a displacing agent is available;

(iii) the high affinity character of polymer adsorption: as long as the surfaces are not saturated, there is practically no free macromolecule in the solvent.

4.3.3 Polymers at solid/liquid interface: effect of charge content on polyelectrolyte adsorption

Lowering the ionicity of the chain, provides all the intermediate situations between polyelectrolyte and nonionic polymer adsorption. This can be done by changing the degree of dissociation of a weak acid or base (e.g. for polyacrylic acid at different pH [7]) or by preparing copolymers between an uncharged monomer and various proportions of a charged one. Figure 4.4 refers to the adsorption of the cationic copolymers of acrylamide mentioned previously onto negatively charged silica particles [11]. The maximum adsorbed amount Γ_{max} is represented as a function of the ionicity for 3 different molecular weights. First of all it must be specified that in this peculiar case, the uncharged unit does not adsorb onto the surface. For any molecular weight, the adsorption increases with the ionicity of the macromolecules, up to $\tau = 1\%$ (because the polymer/surface affinity increases). It also increases with their molecular weight for ionicities lower than 15% because, for such charge densities, the chains are rather flexible (see Figure 4.2), and can form larger and larger loops in solution. When the ionicity of the copolymer increases further, lateral electrostatic repulsions

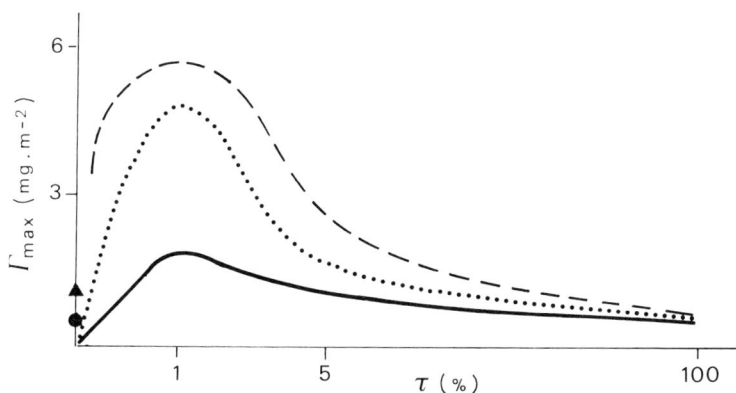

Figure 4.4 Maximum adsorbed amount on silica beads as a function of the chain charge content. For cationic copolymers of acrylamide with molecular weight $M_w = 3 \times 10^6$ (– – –), 5×10^5 (· · · ·), and 10^5 (——). Redrawn from Ref. 12. For PEO with $M_w = 2 \times 10^6$ (▲) and 3×10^4 (●).

grow in the adsorbed layer and the adsorbed amount decreases; simultaneously, molecular weight effects level off, indicating a limitation of the loop size with the increasing stiffness of the chains. The shape of the curves on Figure 4.4 has been predicted and observed for various systems [7,8,12–14].

For the sake of comparison, Γ_{max} values for nonionic poly(ethylene oxide), (PEO) [15, 16] are also reported on Figure 4.4 and are much lower than for low-charged copolymers of this kind, indicating a quite different interfacial conformation between the two types of macromolecules. In the former, any monomer unit can be adsorbed but with a low adsorption energy whilst the latter displays only a few strongly adsorbed cationic segments, the unadsorbed acrylamide units belonging to large loops extending far from the surface. In fact, such a picture is predicted [17] for random copolymers, each time there is a small proportion of one of the monomers displaying a much stronger adsorption energy than the other one. A consequence is that adsorbed layers are much thicker for low charged copolymers than for nonionic homopolymers [12–16] (Figure 4.5). As for Γ_{max}, the M_w dependence of δ disappears at high charge content;

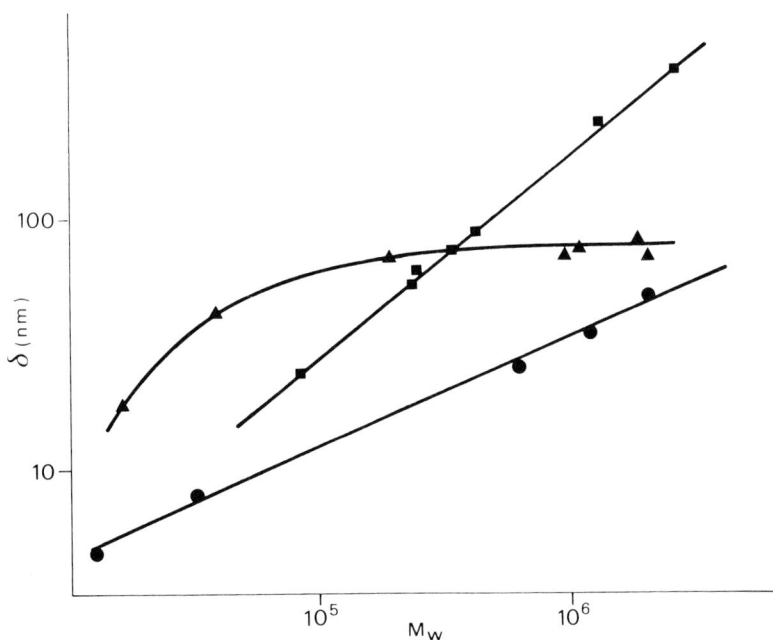

Figure 4.5 Hydrodynamic layer thicknesses as a function of molecular weight for cationic copolymers with $\tau = 1\%(\blacksquare)$, $100\%(\blacktriangle)$. Redrawn from Ref. 12 and for PEO ($\bullet$). Particles are polydisperse silica beads of mean radius 35 nm.

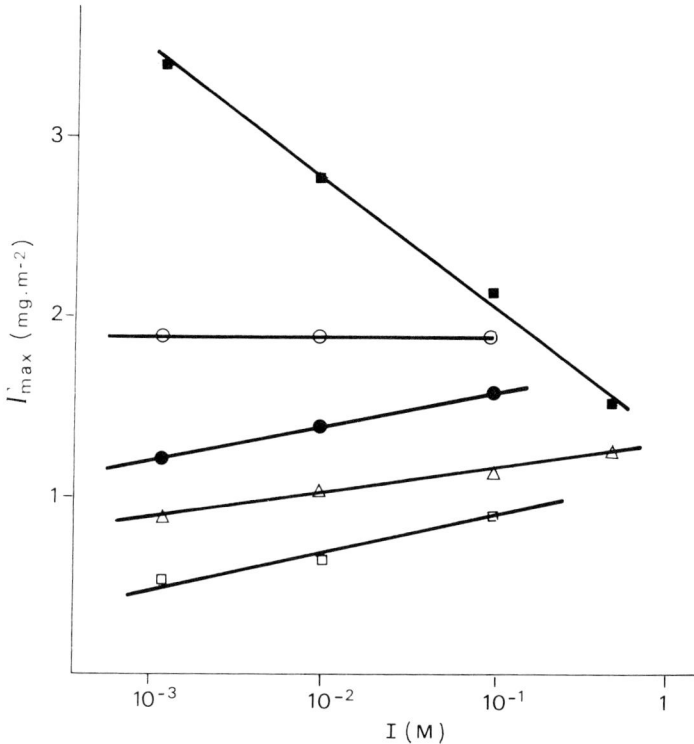

Figure 4.6   Maximum adsorbed amount as a function of ionic strength for cationic copolymers of acrylamide adsorbed onto sodium montmorillonite. $\tau = 30\%$, $M_w = 6 \times 10^5$ ($\square$); $\tau = 13\%$, $M_w = 1.8 \times 10^6$ ($\triangle$); $\tau = 5\%$, $M_w = 10^6$ ($\bullet$); $\tau = 1\%$, $M_w = 1.3 \times 10^6$ ($\blacksquare$); $\tau = 0\%$, $M_w = 3 \times 10^6$ ($\bigcirc$). Redrawn from Ref. 5.

however, layers are far from being as flat as expected, probably because the stiffness of the chains imposes a minimum size for the loops (the order of magnitude of which is about twice the persistence length).

The influence of the ionic strength is rather complex since it depends on the relative charge densities of the chains and of the surface, giving rise to 'screening-enhanced' or 'screening-reduced' adsorption regimes [8]. The overall effect results from the balance between the screening of segment/segment lateral repulsions (which enhances adsorption), the screening of segment/surface attractions (which reduces the proportion of trains) and also the eventuality of a nonelectrostatic polymer/surface interaction. Figure 4.6 shows the evidence of both regimes for cationic polyacrylamides of various charge densities when they are adsorbed onto sodium montmorillonite, a substrate of rather high surface charge density [5].

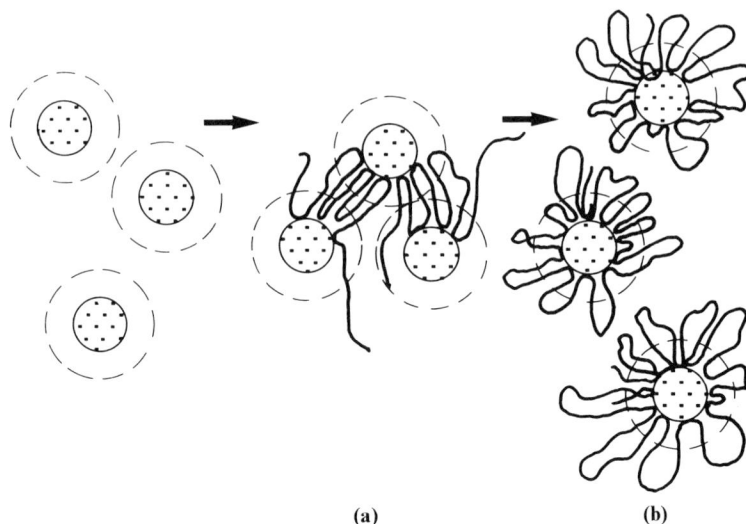

(a)   (b)

Figure 4.7   Effect of the adsorption of a nonionic polymer of high M_w in good solvent conditions on the stability of a colloidal suspension. (a) Unsaturated surfaces: interparticle bridging; (b) saturated surfaces: steric stabilisation. Dashed lines represent the range of the electrostatic repulsions.

4.4   Effect of nonionic polymers on suspension stability: bridging flocculation and steric stabilisation

The adsorption of nonionic polymers on colloidal particles has no important effect on the pre-existing interparticle van der Waals attractions or electrostatic repulsions, but additional interactions between opposite surfaces take place through the adsorbed layers [18]. If surfaces are completely saturated with adsorbed macromolecules, they are kept apart in good solvent by the osmotic repulsions between polymer segments. If the surfaces are unsaturated, they gain adsorption free energy by sharing polymer strands, favouring bridging flocculation thanks to the formation of large flocs. The different steps of uncharged polymer adsorption are illustrated on Figure 4.7. In what follows, it is assumed, of course, that there is no competition between the polymer and a displacing agent (ion or molecule) for the adsorption on the surface.

4.4.1   *Bridging flocculation*

The efficiency of polymer bridging is closely connected with the solvency of the medium and the molecular weight of the polymer. Higher molecular weights permit easier formation of bigger aggregates whereas a good solvent ensures a maximum of extension for the loops and tails in the solution. In aqueous media, polymer bridging will be all the more difficult since electrostatic repulsions have to be overcome, i.e. since the ionic strength is lower; for a given salinity, there is a minimum molecular weight under which the polymer is no longer a flocculant (Figure 4.8). The addition of an indifferent salt at lower concentrations than the c.c.c. generally affects neither the nonionic polymer conformation, nor its adsorption mode, but greatly favours bridging flocculation since it decreases the range of electrostatic repulsion (Figure 4.8). So, the critical M_w to induce flocculation is lower when the ionic strength is increased whereas the aggregate structure becomes more compact [19].

Since restabilisation of the suspension occurs in an excess of added polymer, an optimum flocculation concentration (o.f.c.) can be found, for which the efficiency of bridging is maximum: if the polymer concentration is lower than the o.f.c., the number of interparticle bridges is not sufficient to pick up all particles; conversely, these bridges are more and more difficult to establish when the surface coverage increases because of the occurrence of

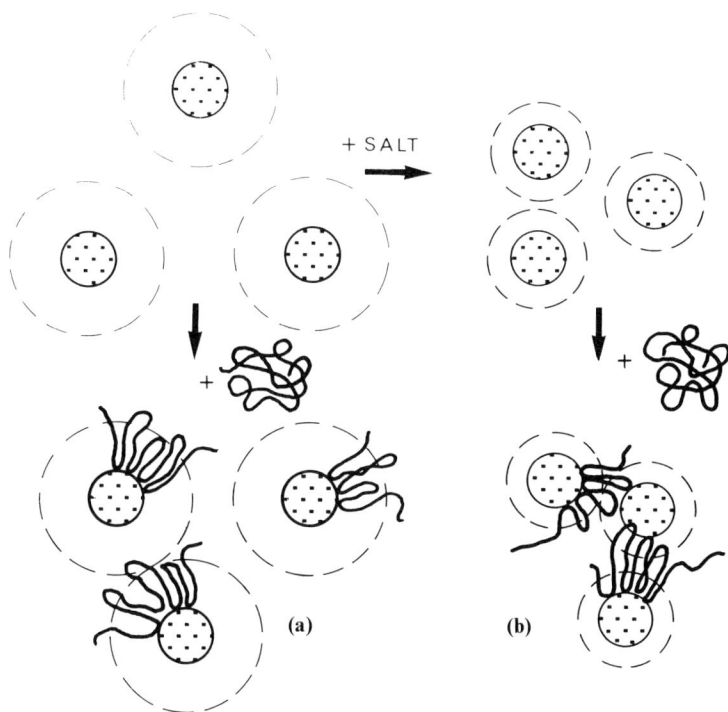

Figure 4.8 Influence of molecular weight and ionic strength on interparticle bridging flocculation. (a) M_w is too low to induce flocculation; (b) at higher I, the same polymer becomes a flocculant. Dashed lines represent the range of the electrostatic repulsions.

osmotic repulsions next to the surface. Various experimental determinations of the o.f.c. can be made by following supernatant clarification, volume of sediment, rate of settling, etc. [3].

From the dynamic point of view, bridging flocculation is a non-equilibrium process. Even at the o.f.c., the method of mixing the polymer and particles is crucial to obtain a homogeneous mixture and to avoid local overdoses that induce partial restabilisation of some particles while there is a lack of flocculant for other ones. The origin of such drawbacks can be found in the irreversibility of polymer adsorption. Moreover, the kinetics of bridging is even more complex since the overall result is governed by the relative rates of polymer adsorption and reconformation at the interface, and of floc formation by collisions between partially covered particles [20,21]. While adsorption and collision times may be approximated with second order reaction kinetics, it is very difficult to estimate the reconformation rate. However, this is of prime importance for flocculation since, at low coverage, adsorption layers can relax and become flattened on the available uncovered surface. It is therefore very important that collisions take place before reconformation in order to form a sufficient number of bridges: this is one of the reasons why bridging flocculation can require a minimum particle concentration to operate [15]. Equilibrium bridging can occur at coverages high enough to prevent fast reconformation or in the presence of salt, when the loop size is less critical for efficient flocculation.

As mentioned above, stirring provides good homogenisation; it accelerates adsorption and flocculation, but for big particles, interparticle collisions are favoured to the detriment of adsorption, resulting in a time-lag between polymer addition and floc formation. More complex effects of breakage and re-formation of flocs are also observed when suspensions are sheared: for instance, in paper manufacture, turbulences are generally used to minimise the unwanted fibre flocculations.

4.4.2   *Steric stabilisation and incipient flocculation*

At high polymer coverage, steric stabilisation occurs thanks to the repulsive interactions between adsorbed layers in good solvent. The steric interaction energy, V_S arises from interpenetration plus compression of the chains in the overlap region; it contains a mixing contribution V_{mix} (which reflects segment/solvent interaction) and an elastic contribution V_{el} (which corresponds to the loss of configurational entropy of the chains due to interpenetration). Although this separation is rather artificial since the two contributions are not independent, their additivity has been derived [18] and represents a good pragmatic approach to reality:

$$V_S = V_{mix} + V_{el}$$

The second term is of course always repulsive, whereas the first one can become attractive in worse than Θ solvents, leading to weak, reversible flocculation ('incipient flocculation'). All these mechanisms have been reviewed in detail elsewhere [18].

From a practical point of view, a good stabilisation requires strong adsorption with layers thick enough to prevent van der Waals attractions, and dense enough to minimise interpenetration. In addition, full coverage is required to avoid bridging and a good solvent to prevent incipient flocculation. These are in fact conflicting conditions since for homopolymers or random copolymers, fluffy layers with only a few segments in the trains are displayed in good solvents. This problem can be circumvented by using block or comb-grafted copolymers: one sequence, insoluble in the solvent provides strong adsorption (anchor), while the other one is chosen to be in good solvent conditions and so ensures the stabilisation (buoy). The architecture of the copolymer (relative sizes of the blocks, proportion and length of the grafts towards the backbone) has to be optimised in order to obtain a final brush-like conformation [22]. In the case of concentrated dispersions, the thickness of the layer has to be optimised too. In order to be easily processed, such systems must actually display a viscosity as low as possible with the highest solid content, so it is important to minimise the volume of the objects, thus to find the minimum layer thickness (i.e. the minimum M_w) that prevents them from aggregation [23]. A great variety of block and graft amphiphilic copolymers have been developed and are currently used as dispersants in formulations for ceramics, cements, pigments or coatings [24, 25].

4.5 Flocculation by polyelectrolytes

4.5.1 *Oppositely, highly-charged polyelectrolytes*

Considering first the adsorption of a polyelectrolyte of high ionicity onto an oppositely charged surface from a salt-free aqueous solvent, the different steps are summarised in Figure 4.9 for a positively charged polyelectrolyte and a negatively charged surface.

Despite the strong segment/surface attraction, polymers cannot be completely flattened on surfaces because the distances between charged monomeric units are generally much shorter than the distances between charged groups on the surface. So, unsaturated surfaces display positively charged patches covered with macromolecules, together with bare negatively charged zones. In such conditions, flocculation occurs through electrostatic attractions between positively and negatively charged domains [3], the best result being obtained when the overall charge of the particle is cancelled. Again, an optimum flocculation concentration can be found,

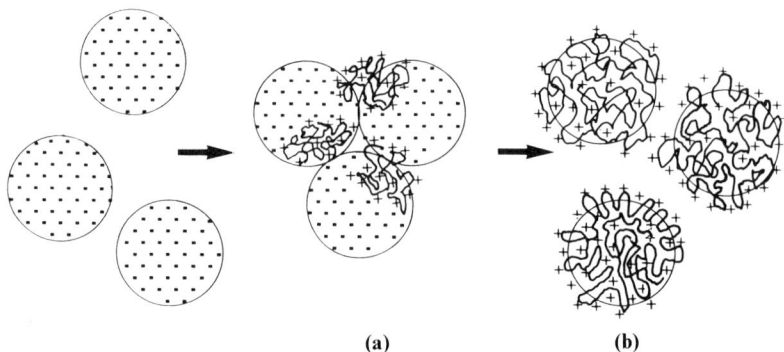

Figure 4.9 Effect of the adsorption of an oppositely charged polyelectrolyte on the stability of a colloidal suspension: (a) Charge neutralisation flocculation; (b) electrosteric restabilisation.

which is governed by the number of charges brought by the adsorbed polymer. Thus, contrary to the bridging situation, for an exclusively charged neutralisation mechanism, the o.f.c. does not depend on the polymer molecular weight and the best clarification of the supernatant occurs together with the charge inversion of the particles (as attested by simultaneous measurements of zeta potential and supernatant turbidity [13,26], an example of which is reported on Figure 4.10).

For saturated surfaces, the overall charge of the particles is reversed and the suspension becomes electrosterically stabilised.

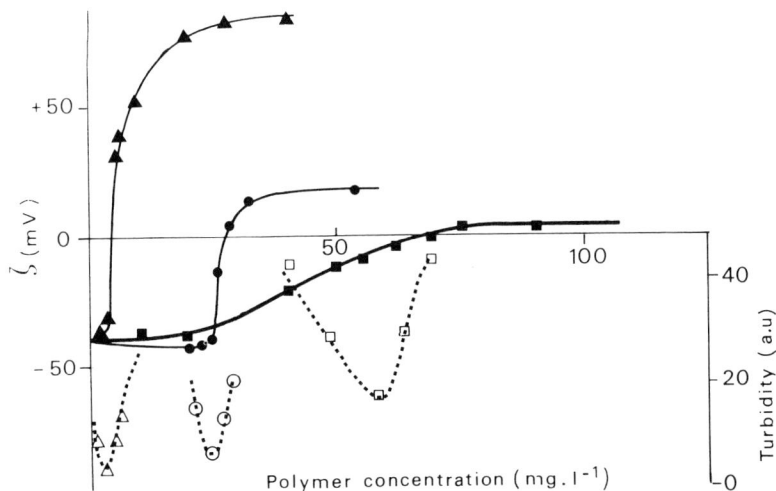

Figure 4.10 Zeta potential (filled symbols) and supernatant turbidity (open symbols) as a function of added polymer concentration for cationic copolymers of acrylamide adsorbed onto sodium montmorillonite: $\tau = 100\%$ (triangles), 13% (circles) and 1% (squares).

4.5.2 *The mixed situations*

In practical conditions, flocculation seldom proceeds through solely bridging or charge neutralisation: both mechanisms often operate simultaneously in combination with a contribution of double-layer compression accounting for the ionic strength of the medium.

The percentage τ of charged units on the chain has a definite influence on the flocculation mechanism [13,27]. If $\tau > 20\%$, the charge neutralisation mechanism dominates with no molecular weight dependence and overall compensation of the particle charge at the o.f.c., whereas for decreasing charge degrees, the o.f.c. is observed for lower and lower polymer concentrations than the isoelectric point (Figure 4.10), and becomes more and more molecular weight dependent (Figure 4.11), which means an increasing bridging contribution to the flocculation mechanism. In the same way the microscopic structure of the flocs evolves from self-similar fractal aggregates to clusters displaying short range liquid-like order (because electrostatic repulsions are not fully compensated) with large scale heterogeneities [28]. These results are in good correlation with the macroscopic floc properties [27].

With polyelectrolytes, the influence of ionic strength is often more complicated than simply making flocculation easier by diminishing the

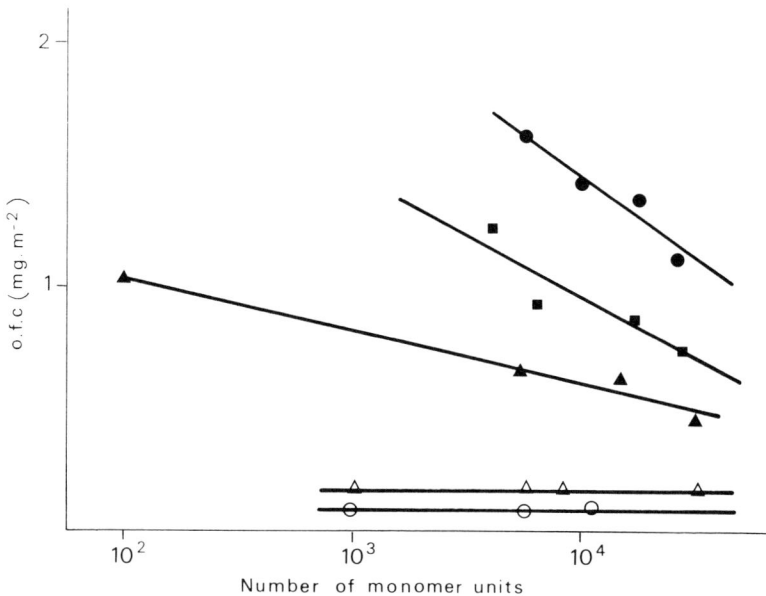

Figure 4.11 Optimum flocculation concentration as a function of chain length for cationic copolymers of acrylamide $\tau = 100\%$ (O), 30% (△), 13% (▲), 5% (■) and 1% (●). Redrawn from Ref. 13.

range of surface/surface electrostatic repulsions. The simultaneous
screening of segment/segment and segment/surface interactions promotes
loop and tail conformation, giving rise to molecular weight effects, even for
high charge content polyelectrolytes. In the 'screening-reduced' adsorption
regime, i.e. for low-charged surfaces and/or chains, high salt concentrations
can even be unfavourable to flocculation in the absence of any nonelectrostatic polymer/surface interaction since the polymer can be
completely displaced from the surface. Practically, increasing the ionic
strength reduces the concentration at which the flocs initially appear and
broadens the flocculation domain [5]. Adsorption of polyelectrolyte chains
bearing charges of the same sign as the surface generally does not induce
efficient flocculation, at least at the macroscopic scale [16] but can be
enhanced by adding a selectively adsorbed species with an opposite charge
[25].

4.5.3   *Flocculation by branched polyelectrolytes and microgels*

In all the cases previously examined, the macromolecular flocculant is
considered as a flexible linear chain. So, in spite of the formation of loops
and tails, each charged polymer unit is able, at a given moment, to be in
direct interaction with a charge of opposite sign located at the particle
surface.

The method of synthesis of very high molecular weight polymers,
frequently used in flocculation treatments (M_w larger than 5 millions),

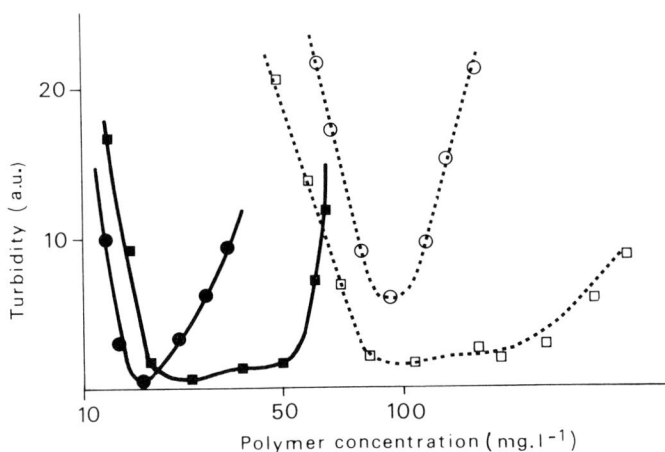

Figure 4.12   Turbidity as a function of polymer concentration for home-prepared (circles) and
industrial (squares) cationic copolymers of acrylamide with $\tau = 100\%$ (full lines) and 5% (dashed
lines). Note the logarithmic scale for polymer concentration.

induces a tendency to branching and even to produce microgels. Consequently non linear chains are often found in commercial flocculants.

As far as the rigidity of the sterically hindered macromolecular species does not prevent the global neutralisation of the oppositely charged particles, no specific effect is observed. However when, under experimental conditions, the size of the polyelectrolyte microgels is smaller than the range of the electrostatic forces in the system (typically κ^{-1}), a fraction of the charges appears as unefficient or poorly efficient. The optimum of flocculation is smoothed and, as a part of the flocculant is unemployed, the o.f.c. is increased (see Figure 4.12). This effect is strongly enhanced by increasing the size of the polyelectrolyte gel [29] which finally induces heteroflocculation as observed when mixing two suspensions of solid particles with oppositely charged surfaces.

4.6   Conclusions

This chapter has described the most general mechanisms of flocculation by adsorbed polymers in relation to their adsorption properties. The major roles of the molecular weight and ionicity of the chains, together with the ionic strength of the medium have been underlined. In many cases, the efficiency of polymeric flocculants compensates for their rather high cost and difficulties of processing since large and shear-resistant aggregates are obtained for very low doses of reagent. However, it must be emphasised that adsorption and flocculation properties can be greatly modified in the presence of species which compete to interact either with the chains or with the surface. For instance when H^+ and OH^- ions are involved in such reactions (i.e. when they are potential determining ions for the surface [15,16] and/or when the charged units are weak acids or bases [14]), thus care must be taken to master the variation of any external parameter that can change the state of the system. So, in practical situations, it is important in the same way to check the compatibility of the polymer with all the constituents of the formulation. Those additional interactions allow modulation of the aggregation level and are often used to develop specific applications such as binders or reversible and selective flocculation processes.

References

1. Israelachvili, J.N., *Intermolecular and Surface Forces*, Academic Press (1991).
2. Fowkes, F.M. and Pugh, R.J., Steric and electrostatic contributions to the colloidal properties of nonaqueous dispersions, in *Colloid Properties of Nonaqueous Dispersions* ACS 240 (1984), 331–54.

3. Ives, K.J., *The Scientific Basis of Flocculation*, NATO Advanced Series E n° 27, Alphen aan den Rijni (1978), The Netherlands.
4. Parfitt, G.D. and Rochester, C.M., *Adsorption from Solution at the Solid/Liquid Interface*, Academic Press (1983).
5. Durand, G., Lafuma, F. and Audebert, R., 'Adsorption of cationic polyelectrolytes at clay-colloid interface in dilute aqueous suspensions: Effect of the ionic strength of the medium.' *Progress in Colloid & Polymer Science*, **266** (1988), 278–82.
6. de Gennes, P.G., 'Polymers at interfaces, a simplified view.' *Advances in Colloid and Interface Science*, **27** (1987), 189–209.
7. Blaakmeer, J., Böhmer, M.R., Cohen-Stuart, M.A. and Fleer, G.J., 'Adsorption of weak polyelectrolytes on highly charged surfaces. Poly(acrylic acid) on polystyrenee latex with strong cationic groups.' *Macromolecules*, **23** (22) (1990), 2301–09.
8. Van de Steeg, H.G.M., Cohen-Stuart, M.A., de Keizer, A. and Bijterbosch, B.H., 'Polyelectrolyte adsorption: A subtle balance of forces.' *Langmuir*, **8** (1992) 2536–46.
9. Lecourtier, J., Lee, L.T. and Chauveteau, G., 'Adsorption of polyacryamides on siliceous minerals.' *Colloids and Surfaces*, **47** (1990), 219–31.
10. Järnström, L. and Stenius, P., 'Adsorption of polyacryline and carboxy methyl cellulose on kaolinite: Salt effects and competitive adsorption.' *Colloids and Surfaces*, **50** (1990), 47–73.
11. Denoyel, R., Durand, G., Lafuma, F. and Audebert, R., 'Adsorption of cationic polyelectrolytes onto montmorillonite and silica: Microcalorimetric study of their conformation.' *Journal of Colloid and Interface Science*, **139** (1990), 281–90.
12. Wong, T.K. and Audebert, R., 'Adsorption of cationic copolymers of acrylamide at the silica–water interface: Hydrodynamic layer thickness measurements.' *Journal of Colloid and Interface Science*, **121** (1) (1988), 32–41.
13. Durand-Piana, G., Lafuma, F. and Audebert, R., 'Flocculation and adsorption properties of cationic polyelectrolytes towards Na-montmorillonite dilute suspensions.' *Journal of Colloid and Interface Science*, **119** (2) (1987), 474–79.
14. Vaslin-Reimann, S., Lafuma, F. and Audebert, R., 'Reversible flocculation of silica suspensions by water-soluble polymers.' *Colloid and Polymer Science*, **268** (1990), 479–83.
15. Lafuma, F., Wang, K. and Cabane, B., 'Bridging of colloidal particles through adsorbed polymers.' *Journal of Colloid and Interface Science*, **143** (1) (1991), 9–21.
16. Lafuma, F., unpublished results.
17. Marques, C.M. and Joanny, J.F., 'Adsorption of random copolymers.' *Macromolecules*, **23** (1990), 268–76.
18. Napper, D.H., *Polymeric Stabilization of Colloidal Dispersions* Academic Press, London (1983).
19. Wong, K., Lixon, P., Lafuma, F., Lindman, P., Aguerre-Charriol, O. and Cabane, B., 'Intermediate structures in equilibrium flocculation.' *Journal of Colloid and Interface Science*, **153** (1) (1992), 55–72.
20. Dickinson, E. and Eriksson, L., 'Particle flocculation by adsorbing polymers.' *Advances in Colloid and Interface Science*, **34** (1991), 1–29.
21. Adachi, Y., Cohen-Stuart, M.A. and Fokkink, R., 'Dynamic aspects of bridging flocculation studied using standardized mixing.' *Journal of Colloid and Interface Science*, **167** (1994), 346–51.
22. Fleer, G.J. and Scheutjens, J.N.H.M. 'Block copolymer adsorption and stabilization of colloids.' *Colloids and Surfaces*, **51** (1990), 281–98; Wu, D.T., Yokoyama, A. and Setterquist, R.L., 'An experimental study on the effect of adsorbing and non-adsorbing block sizes on diblock copolymer adsorption.' *Polymer Journal*, **23** (5) (1991), 709–14; Israëls, R., Leermakers, F.A.M. and Fleer, G.J., 'Adsorption of charged block copolymers: Effect on colloidal stability.' *Macromolecules*, **28** (1995), 1626–34.
23. Corradi, A.B., Manfredini, T., Pellacani, G.C. and Pozzi, P., 'Deflocculation of concentrated aqueous clay suspensions with sodium polymethacrylates.' *Journal of the American Ceramic Society*, **77** (2) (1994), 509–13.
24. Jakubauskas, H.L., 'Use of A–B block polymers as dispersants for non-aqueous coating systems.' *Journal of Coatings Technology*, **58** (736) (1986), 71–82; Rudolph, J., Patzsch, J., Meyer, W.M. and Wegner, G., 'The interaction of acrylic diblock-copolymers with aluminium oxide surfaces and their application for ceramic powder processing.' *Acta Polymerica*, **94** (1993), 230–37.

25. Somasundaran, P. and Yu, X., 'Flocculation/dispersion of suspensions by controlling adsorption and conformation of polymers and surfactant.' *Advances in Colloid and Interface Science*, **53** (1994), 33–49.
26. Mabire, F., Audebert, R. and Quivoron, C., 'Flocculation properties of some water-soluble cationic copolymers toward silica suspensionsss: A semiquantitative interpretation of the role of the molecular weight and cationicity through a patchwork model.' *Journal of Colloid and Interface Science*, **97** (1) (1984), 120–36.
27. Wang, T.K. and Audebert, R., 'Flocculation mechanisms of a silica suspension by some weakly cationic polyelectrolytes.' *Journal of Colloid and Interface Science*, **119** (20) (1987), 459–65.
28. Cabane, B., Wong, K., Wang, T.K., Lafuma, F. and Duplessix, R., 'Short range order of silica particles bound through adsorbed polymers layers.' *Colloid and Polymer Science*, **266** (1988), 101–104.
29. Ferrer, E. and Audebert, R., Amphiphilic cationic polymers as both flocculants and depollutants, in *Chemical Water and Wastewater Treatment*, eds R. Klute and H. Hahn, Springer Verlag, Berlin (1992), 33.

5 Retention aids

D. HORN and F. LINHART

5.1 Introduction

Papermaking is basically a filtration process. The paper machine wire can be regarded as being a continuous filter on which a proportion of the solids in the stock is retained. The unretained solids drain through the wire along with the majority of the liquid to form the white water. The separation of these two phases, i.e. the drainage process, must be sufficiently advanced when the freshly formed web of paper leaves the wire in order to ensure that no web breaks occur. The rate at which the two phases are separated dictates the rate at which paper can be produced. Drainage rates have a large bearing on the economics of papermaking.

They can be controlled by adjusting the mesh of the wire or the pressure drop between the upstream and downstream sides of the wire. High pressure drops and, especially, a coarse mesh have the result that there is less separation between the two phases, retention is lower and the papermaking process becomes more involved and less economical. Paper quality is often impaired. Less fines and fillers are retained on coarse-mesh wires, but these are important for ensuring good printability. The majority of the white water is returned to the stock, with the result that the concentration of fines and fillers gradually increases. These will eventually find their way into the paper, but this can only be accomplished at the cost of high solids in the white water with the attendant problems of deposits, lower paper quality and difficulties in treating effluent.

A distinction is made between first-pass or one-pass retention and total retention. Total retention is defined as the ratio of the mass of the paper leaving the machine and the mass of the raw materials employed, expressed in percent. First-pass retention is the ratio between the solids retained on the wire and the solids in the headbox.

It was discovered that certain chemical substances could be used to increase the retention of fines and fillers and, at the same time, to accelerate drainage. The two phases are separated more efficiently, with the result that paper quality is improved, the productivity of the machine is greater and a quicker return on capital can be achieved. There is hardly a single paper mill in Central, Western or Southern Europe that does not use retention aids.

Recently, retention aids have begun to be used in the production of wood-containing printing papers that contain no mineral fillers – Northern European and North American newsprint are typical examples. World consumption of synthetic retention aids totalled *c.* US$ 600 million in 1986.

5.2 The chemical nature of retention aids

Retention aids can be grouped together as follows:

- Inorganic retention aids
- Retention aids based on natural organic raw materials
- Synthetic, water-soluble organic polymers

5.2.1 *Inorganic retention aids*

The most important inorganic retention aids are the aluminium salts. Aluminium sulphate, erroneously referred to in the paper industry as alum, has been in use ever since rosin was first used to size paper. One of its functions is to provide an increase in retention, but its main use is to make it possible for rosin to be effective as a size. This was the reason that it was first introduced into papermaking. It improves the runnability of paper machines by fixating resinous pitch and other sticky substances to fibres and fillers, thus preventing them from being deposited. Unlike most other retention aids, the species responsible for improving retention does not form a solution in the 4–8 pH range at which paper is normally produced. It forms a positively charged colloid, whose properties depend on both pH and the nature of the anion in the aluminium salt [1–3]. 'Polyaluminium chloride', a basic aluminium chloride, is increasingly being used as a source of aluminium ions.

Aluminium salts are normally employed in concentrations of between 0.5 and 3%, in terms of the oven-dry stock. Aluminium salts are hydrolysed, which causes a drop in the pH of the paper stock, the extent of which depends on the stock's buffering capacity. The use of cationic aluminium salts entails that the paper is made in acid media. Sodium aluminate is occasionally used, mainly to raise the pH of the stock and to provide a fresh source of colloidal aluminium.

Some inorganic substances are used in combination with other chemicals to aid retention. For instance, silicic acid is used in combination with cationic starch, but the main aim here is to improve web strength [4].

Combinations of alkaline-activated bentonite and nonionic high molar mass polyacrylamides often lead to an increase in retention and drainage rates [5,6]. Bentonite consists predominantly of montmorillonite, a clay mineral which swells in water, but which on its own has only a negligible effect on retention.

5.2.2 *Retention aids based on natural organic raw materials*

Cationic starches [7] are the main substances that fall under this heading. Their main purpose is to improve dry strength, and improved retention is only a welcome side effect. Starch is modified predominantly with 2,3-epoxypropyl-1-trimethylammonium chloride to render it cationic, but chloroethyldimethyl-ammonium chloride is also used for this purpose. Other modified polysaccharides, such as cationic guar gum, have also been classified as retention aids, but the main function of this group of substances is to improve strength.

5.2.3 *Synthetic, water-soluble organic polymers*

The substances normally referred to as retention aids in the paper industry are organic, water-soluble high molar-mass polymers. The main consideration in the design of manufacturing processes is to achieve optimum retention and drainage. The polymers can be anionic, cationic or nonionic, but their molar mass also plays an important part in determining their effectiveness.

Synthetic retention aids can be classified according to their monomers. They are mostly polyacrylamides, polyamines, polyethylenimines, polyamido-amines and polyethylene oxides. These groups of substances are discussed in detail below. Some suppliers have also propagated polyionenes and polypyrrolidinium derivatives as retention aids.

5.2.3.1 *Polyacrylamides (PAM)*.
Polyacrylamides [8–11] are widely used as retention aids in the paper industry. Non-ionic polyacrylamides (structure I) are homopolymers produced from acrylamide with the aid of a free-radical initiator.

Ionic polyacrylamides (structure II, R = charge-carrying group) are either copolymers of acrylamide and other anionic or cationic vinyl monomers, or they are made by modifying nonionic polyacrylamides.

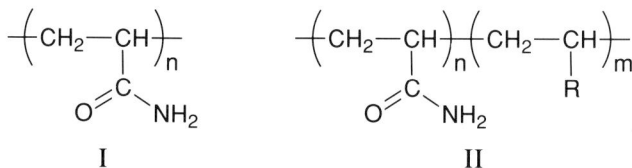

I   II

Cationic polyacrylamides are copolymers of acrylamide and various cationic derivatives of acrylic acid. The most important group of substances are esters of acrylic or methacrylic acid and a dialkylaminoethanol (structure III), such as dimethylaminoethylmethacrylate (R_1, $R_2 = CH_3$; $R_3 = H$ in acidic solution). Other monomers include quaternary ammonium esters synthesised with methyl chloride or dimethyl sulphate, such as acryloylethyl-trimethylammonium chloride ($R_1 = H$; $R_2, R_3 = CH_3$; $X = Cl$).

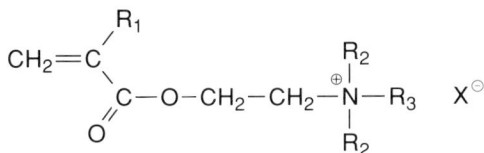

$$CH_2=C \overset{R_1}{\underset{\overset{\displaystyle \parallel}{O}}{\overset{\displaystyle /}{\underset{C}{\diagdown}}}}-O-CH_2-CH_2-\overset{R_2}{\underset{R_2}{\overset{\mid}{\overset{\oplus}{N}}}}-R_3 \quad X^{\ominus}$$

III

Cationic amides prepared from acrylic acid or methacrylic acid are also used, though less frequently, to introduce positive charges into the polyacrylamide chain, e.g. methacrylamidopropyl-trimethylammonium chloride. It must be pointed out that products that are polymerised from 90–100% of cationic monomeric esters are often wrongly termed 'cationic polyacrylamides'.

Cationic polyacrylamides containing 20–70% cationic moieties are normally supplied to the paper industry. The charge densities of polymers range from between c. 0.75 and 3.5 mVal/g, depending on the molar mass of the cationic monomers. Cationic polyacrylamides can also be prepared from the homo-polymer by modifying it with dimethylamine and formaldehyde, as described by Einhorn [11, 12].

Anionic polyacrylamides (structure IV) are copolymers of acrylic acid or sodium acrylate, but they can also be manufactured by the controlled hydrolysis of polyacrylamides up to the required charge density.

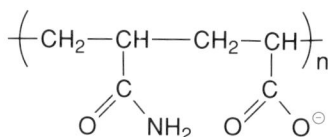

$$\left(CH_2-\underset{\underset{O}{\overset{\diagup}{\underset{NH_2}{\overset{\displaystyle C}{\diagdown}}}}}{\overset{\mid}{CH}}-CH_2-\underset{\underset{O}{\overset{\diagup}{\underset{O^{\ominus}}{\overset{\displaystyle C}{\diagdown}}}}}{\overset{\mid}{CH}}\right)_n$$

IV

The increase in retention provided by modern polyacrylamides depends to a very great extent on their molar mass. Polymers of a given charge density will have a greater effect on retention the higher their molar mass. The polyacrylamides in use today have such high molar masses and their aqueous solutions are so viscous that they can only be conveniently handled at concentrations of less than 2%. It is for this reason that aqueous solutions have become commercially much less important in the last few years. Nowadays, most retention aids based on polyacrylamides are supplied as solids or in the form of water-in-oil emulsions.

Solid products are manufactured by polymerising the monomers in water or in the aqueous phase of a water-in-oil (W/O) emulsion of an immiscible organic solvent. The water and, in some cases, the organic solvent are distilled off after polymerisation. Solid polyacrylamides are supplied in the form of granules or small beads. They dissolve fairly slowly in water, and swell before they dissolve. Care must be taken to ensure that no gelatinous particles find their way into the paper stock, as this can cause web breaks or, at least, pinholes and spots in the paper. Expensive dissolving and filtration equipment is required for solid polyacrylamides, and equipment costs can often outweigh the economic benefits provided by retention aids in small paper mills.

Water-in-oil emulsions are polymerised after dissolving the monomers in the aqueous phase of a W/O emulsion of a high-boiling paraffin. The polyacrylamide is dispersed in the form of small particles, which are swollen due to solvation with water. These dissolve very quickly when sufficient water is added, but a wetting agent is required to bring the polyacrylamide into the aqueous phase, and this can be added during manufacture or dilution. The W/O emulsions currently on the market have a polymer content of 25–50%. They are an economically viable alternative to solid products.

5.2.3.2 *Polyethylenimine (PEI).* Polyethylenimines [13] and their derivatives have long been employed to improve drainage and retention [14–16]. The polymer (structure V) is manufactured from aziridine (ethylenimine), which is highly reactive, by a ring-opening polyaddition reaction in aqueous solution in the presence of an acid catalyst.

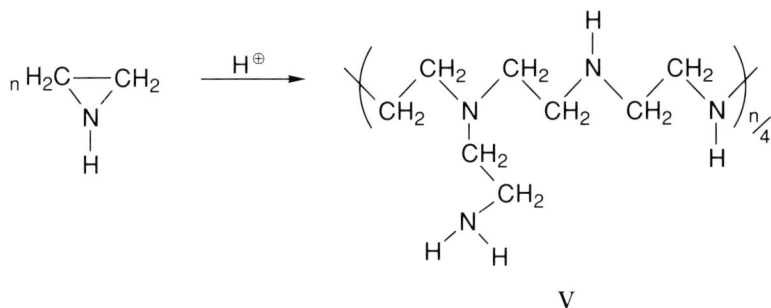

V

The finished product contains no residual monomers. Its molar mass is controlled by the conditions under which the reaction takes place and the nature and quantity of acid initiator employed. As with other retention aids, their effectiveness increases in line with their molar mass, and the aim is always to obtain a product with the highest molar mass possible. The problem here is that polyethylenimine chains are branched, and there is a danger that an insoluble gel will be obtained if the molar mass is too high. Because it is branched, polyethylenimine contains primary, secondary and tertiary amino groups, approximately in the ratio 1:2:1 [17].

Aziridine has a relative molar mass of 43, which is low, and this is the reason why polyethylenimine is the polyelectrolyte with the highest charge density (23.3 mVal/g) in aqueous solution in its fully protonated form. Of course, this figure decreases as pH increases [13].

High molar mass polyethylenimines are supplied commercially in the form of highly alkaline aqueous solutions with a concentration of between 30 and

50%. They consist of polymer, water and residual traces of initiator.

Products with a higher molar mass can be derived from low-molecular-weight polyethylenimines by crosslinking. Polymers identical to homopolymers in terms of their structural units, but which contain additional chloride ions, can be prepared by using 1,2-substituted electrophilic ethane derivatives such as 1,2-dichloroethane as a crosslinking agent.

PEI derivatives with additional functional groups, higher molar masses and lower charge densities have been developed in order to extend the range of properties that polymer structures of this type display. Polyamidoamines, i.e. polycondensation products of adipic acid, form the basis of these polymers, which can be crosslinked to increase their molar mass. Alkaline solutions are obtained, which are partially neutralised and supplied commercially at concentrations of 15–20%. In spite of their being relatively highly concentrated, their viscosities are low, and they are easy to handle and readily miscible with water.

5.2.3.3 *Polyamines*. The polyamines employed as retention aids in the paper industry are usually polycondensation products of amines and short-chain crosslinking agents, which normally contain chlorine. The simplest representatives of this group of substances are the condensation products of ethylenediamine and dichloroethane [18].

Oligomers such as diethylenetriamine, triethylenetetramine and their higher homologues or ammonia itself can be used in place of ethylenediamine. It can be assumed that polymers of this type are similar to polyethylenimine in terms of their structural units. Chloride ions are liberated in substantial quantities during the polycondensation reaction. Products with a molar mass high enough to permit their use as retention aids are supplied commercially in the form of 20–40% solutions. Like polyethylenimines, polyamines can be combined with other water-soluble polymers to boost their effectiveness. The polyamines can be crosslinked with reactive prepolymers formed from polyamidoamines (see below), with the aid of bifunctional compounds such as epichlorohydrin, to form high molar mass polycondensates.

Polyamines, or mixtures of polyamines and quaternary polyammonium compounds, can also be obtained in condensation reactions between dimethylamine and epichlorohydrin. Other possible reagents are ammonia, methylamine, ethylenediamine, diethylenetriamine or other polyfunctional amines on the one hand, and epichlorohydrin or aliphatic dichlorides on the other hand. Many different combinations are possible, but little is known about the polymers' exact chemical compositions or their physical properties. Most of the products of this type available commercially are supplied at a concentration of 40–50% and their viscosities are relatively low, which would indicate that their molar mass is not particularly high. In Europe, products of this type are principally used to fixate anionic polyacrylamides and other ionic polymers, rather than as retention aids.

5.2.3.4 *Polyamidoamines.* Water-soluble resins can be obtained by condensing adipic acid and diethylenetriamine at high temperature, in the course of which water is eliminated. The polyamidoamines thus obtained can be caused to react with an excess of epichlorohydrin to form polyamidoamine-epichlorohydrin resins, which are employed as neutral wet-strength additives. If only small amounts of epichlorohydrin are employed or a different crosslinking agent is used, a high molar mass compound is obtained which can be used as a retention aid [19].

Other monomers, such as caprolactam, dimethyladipate or triethylenetetramine or higher oligomeric diamines can be used in place of the monomers listed above, or in addition to them. The charge densities of cationic polyamidoamines of this type are in the range 3–5 mVal/g, and they are supplied commercially at concentrations of between 20 and 30%.

5.2.3.5 *Other cationic polymers.* A number of polymers containing pyrrolidinium groups [20,21], prepared from diallyldimethylammonium chloride (structure VI) have been shown to have an effect on retention, and they may well have been used specifically for this purpose, but the main function of polyDADMAC, supplied commercially at concentrations of 30–50%, is to fixate anionic oligomers and polymers.

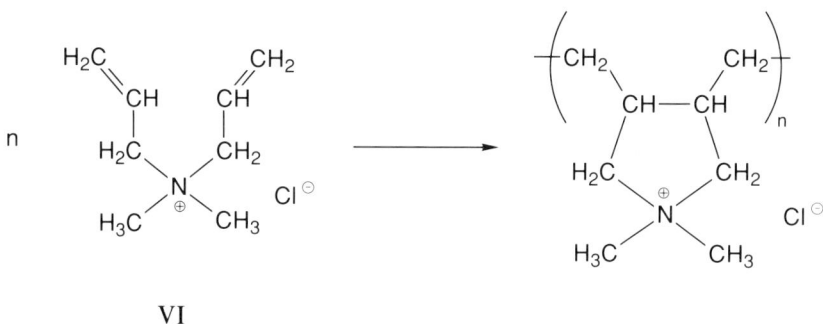

VI

Similarly, polyionenes [22], which are condensation products of alpha, omega-dichloroalkanes and dimethylamine or alpha, omega-N,N'-tetramethyldiaminoalkanes, are used for much the same purpose.

5.2.3.6 *Polyethylene oxide (PEO).* Polyethylene oxide is a very familiar substance, but it is only recently that it has been in regular use in the paper industry as a retention aid [23–25]. Unlike all other retention aids, only nonionic PEO (structure VII) with a high molar mass is employed. It is polymerised by means of precipitation in the presence of a basic catalyst.

Polyethylene oxides have to have a high relative molar mass if they are to function as retention aids, and a figure of 4 million has been quoted as a minimum. The C–C– and C–O– groups in the monomers form the backbone

$$n \quad \underset{O}{H_2C\!\!-\!\!CH_2} \quad \longrightarrow \quad \left(\!H_2C\!-\!CH_2\!-\!O\!\right)_n$$

VII

of the polymer chains, and very long linear structures are obtained. Even dilute solutions of PEO have a very high viscosity. PEOs are supplied commercially in the solid form only, the form in which they are obtained in the manufacturing process.

5.3 Mode of action of retention aids

Retention aids act by allowing the extent of flocculation in stock suspensions to be controlled. With a few exceptions, the paper fibres, fillers, resinous particles and other components that make up the stock suspension are negatively charged. The continuous aqueous phase varies according to its pH, the concentrations of monovalent and multivalent metal ions it contains, and by the presence of water-soluble ionic or neutral substances extracted from groundwood, chemical pulp or deinked stock. The process water may also contain impurities, notably humic acid.

Retention aids interact with the various different substances in the stock, with the aim of controlling the level of co-flocculation so as to ensure maximum separation of the dispersed and dissolved constituents from the water. It is also important that formation is not impaired.

Figure 5.1 is a schematic representation of the most important means by which ionic retention aids interact with the various components of stock suspensions. Cationic polyelectrolytes adsorb spontaneously on negatively charged surfaces, while anionic polymers have to be induced to adsorb with the aid of polyvalent metal ions or cationic polyelectrolytes. After completion of the elementary stage of adsorption, changes in the conformation of molecules in the adsorbed state lead to specific structures of the adsorbate layer which, in turn, lead to different flocculation mechanisms.

The kinetics of these processes and the final conformations of the polymer molecules are essentially determined by their charge densities and molar masses and by the charge density and morphology of the adsorbent surface. The charge density of polycarboxylates or polyamines with primary, secondary or tertiary nitrogen functions can be modified by adjusting the pH. The diameter of polyelectrolyte molecules can also be varied within wide limits at the same time.

The various substances in the continuous phase are able to enter into specific interactions, by counter-ion binding or polymer-complex formation, which may provide advantages in that polyanion adsorption is induced, but also disadvantages, because polyanions may impair the effectiveness of cationic retention aids.

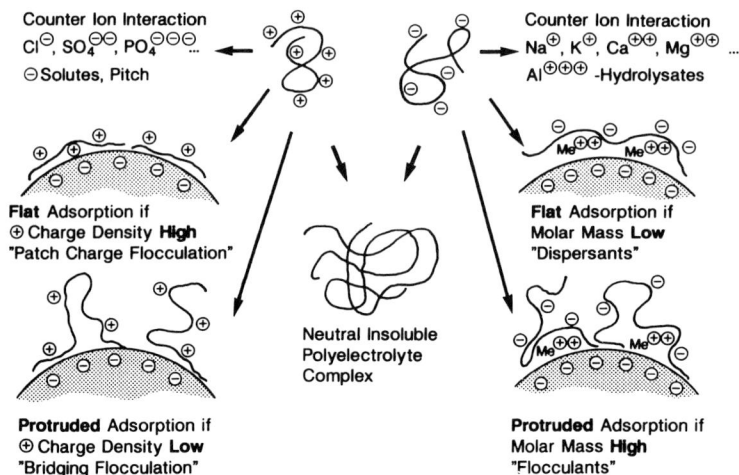

Figure 5.1 Ionic retention aids' modes of interaction with various components of stock suspensions.

5.3.1 *Characterisation of retention aids*

In order to come to an understanding of the mode of action of a retention aid its charge density, molar mass and coil diameter have to be determined. Its adsorption, interaction with metal and polymer counter-ions and, last but not least, flocculation efficacy also have to be quantified. A summary of the most important methods of analysis is given in Table 5.1.

Polyelectrolyte titration is one of the most important methods of characterising ionic retention aids. It has repeatedly been observed that anionic and

Table 5.1 Experimental methods for the characterisation of retention aids in solution and of their mode of action in paper stock suspensions

Property	Experimental method	References
Molar mass	Light scattering	[26]
	Gel-permeation-chromatography	[27]
	Sedimentation-analysis	[28]
Coil diameter	Dynamic light scattering	[29, 30]
	Viscosimetry	[31]
Charge density	Polyelectrolyte titration	[32, 33]
	Potentiometry	[34, 35]
	Conductimetry	[36]
Counter ion interaction	Ion selective potentiometry	[37, 38]
Adsorption at interfaces	Polyelectrolyte titration	[33]
	Electrophoresis	[39, 40]
State of flocculation	Dynamic flocculation testing	[41–44]

cationic polyelectrolytes form stoichiometric polyelectrolyte complexes in dilute aqueous solutions characterised by a 1:1 pair interaction between the positive and negative centres of charge [13, 32, 45–47]. In the most sensitive version, the end point is determined colorimetrically and, if a special photometric titrator is used, a limit of detectability of the order of < 0.05 ppm can be achieved [13, 33].

Polyelectrolyte titration allows the charge densities of retention aids and their concentrations in aqueous solutions to be measured, which in turn allows adsorption isotherms at variable pH values to be determined with great precision. The interaction of retention aids with pitch and other disruptive ionic substances can be quantified in a similar way [33, 48].

5.3.2 Adsorption on interfaces

The adsorption of ionic retention aids on mineral fillers readily corresponds to theoretical principles [49]. The mechanisms of adsorption on chemical pulp and groundwood, however, are more involved, and depend on the type of wood, the means by which the stock is prepared and its freeness. A marked influence is exerted by variations in porosity and in the chemical nature of the

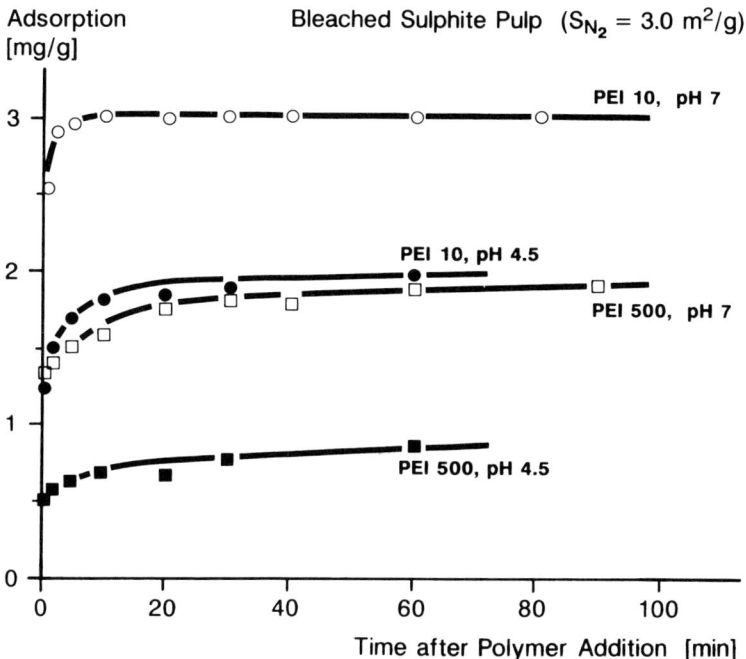

Figure 5.2 Adsorption versus time of PEI 10 and PEI 500 onto bleached sulphite pulp ($S_{N_2} = 3.0\,m^2/g$) at pH 7.0 and pH 4.5 [51]. Cellulose concentration 2 g/dm^3. Level of addition, 0.3%. For PEI specifications see [13].

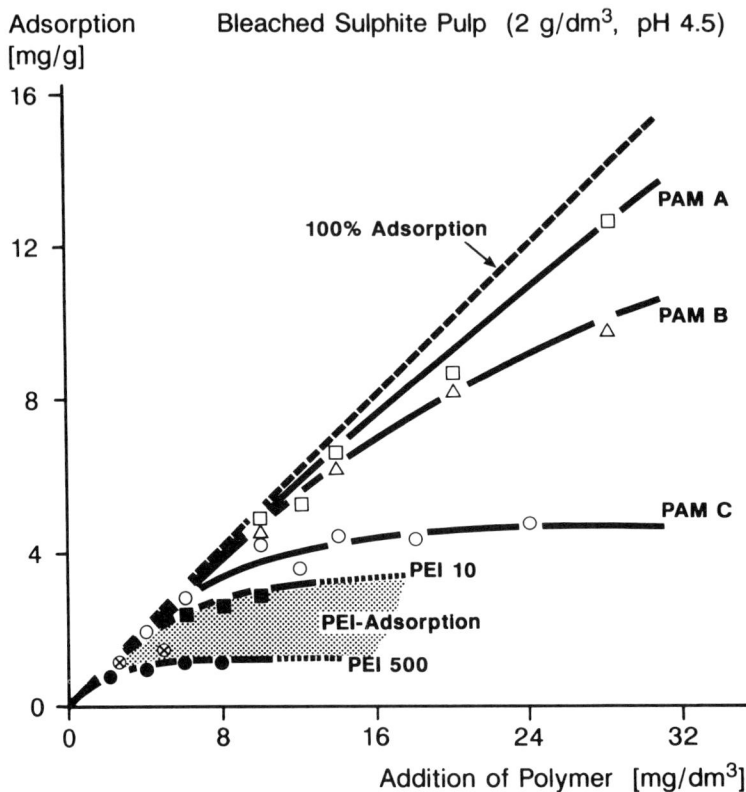

Figure 5.3 Adsorption of polyacrylamides and polyethylenimines of different molar mass onto bleached sulphite pulp. The molar mass of the polyacrylamides varied between 0.2×10^6 (A) and 2×10^6 (C). The investigations with PEI cover molar masses between 400 (PEI 10) and 25 000 (PEI 500). $\otimes$ denotes data points with a modified PEI of 1.8×10^6 molar mass. For further details see [13, 51].

functional groups. In general, it has been observed that adsorption decreases as the molar mass and charge density of the molecule increases [13, 33, 50–54]. An explanation for the decrease in adsorption with increasing molar mass is based on the porous nature of the substrate which limits the access of larger molecules to the internal surfaces of the voids. The accessibility of cellulose is discussed more fully in Chapter 2. Similarly, the intermolecular and intra-molecular repulsion forces between molecules of like charge cause them to occupy a greater area as their charge density increases and, thus, less adsorption takes place. If changes in charge density are caused by changes in pH, concomitant changes in the charge density on the substrate's surfaces and changes in the ionic strength of the dispersion medium may also affect adsorption. The results of these changes can only be determined experiment-ally [13, 51, 55–57].

It is important to note that the above only applies to experiments in which adsorption takes place in equilibrium. Kinetic effects also play an important part under practical conditions [50, 51, 58]. In Figure 5.2, the interaction of pure polyethylenimine with a bleached sulphite pulp is taken as the model: molar mass, charge density (i.e. changes in pH) and adsorption kinetics compete for influence [51]. In Figure 5.3, the adsorption isotherms of highly charged polyethylenimines and weakly cationic modified polyacrylamides with different molar masses are shown in comparison [33, 51]. The charge densities of these polyacrylamides are approximately 1/15 of those of the polyethylenimines. The adsorptive capacities of the polyacrylamides are higher in all cases. This is the first indication that the structures of PAM$^+$ adsorbate layers are basically different to those of polyethylenimines. It is interesting to note that, here too, the polymers' tendency to adsorb increases with decreasing molar mass.

5.3.3 *Electrokinetic effects of polymer adsorption*

Electrokinetic experiments performed on the same system, by measuring the streaming potential or electrophoretic mobilities, have shown that the initially negative charge at the fibre surface decreases in a manner specific to the polymer and the paper stock [13, 33, 51, 59–61]. The quantity of retention aid required for an apparent isoelectric point to be reached is described as the 'cationic demand' specific to the stock suspension [59, 62, 63]. The interfaces become positively charged if more polymer is added [13, 51, 59]. Hence, electrokinetic measurements are useful under practical conditions for determining optimum levels of addition, provided that a means of correlating levels of addition and retention is determined *empirically* in the retention aid/stock system concerned. Attempts to convert electrokinetic measurements such as electrophoretic mobility and streaming potential into the zeta potential in order to make a general prognosis on flocculation, retention or drainage rates have little theoretical basis.

The only case in which it is justified to convert the mobility values into zeta potentials, by means of the Smoluchowski equation [64] for example, is when the potential at the interface is controlled due to interaction with monovalent or bivalent counter-ions within the diffuse double layer. Polyelectrolytes, however, form interfaces whose electrokinetic behaviour is determined by the sum of the portions of the interface that are occupied or unoccupied with retention aid molecules. This is especially true at levels below the saturation point, and the electrokinetic behaviour of structures of this type is very difficult to describe in theoretical terms [65].

5.3.4 *Polymer-induced flocculation*

Apart from the large differences in their adsorption isotherms, extensive experimental work on various pulp/filler systems and retention aids with high,

Patch Charge Model

Polycation:

High Charge Density
Polymer Network
or Chains

F

U_e ⊕

0

⊖

Flocculation

Mobility

Polymer Dosage

Bridging Model

Polycation:

Low Charge Density
High Molar Mass
Chains

F

U_e ⊕

0

⊖

Flocculation

Mobility

Polymer Dosage

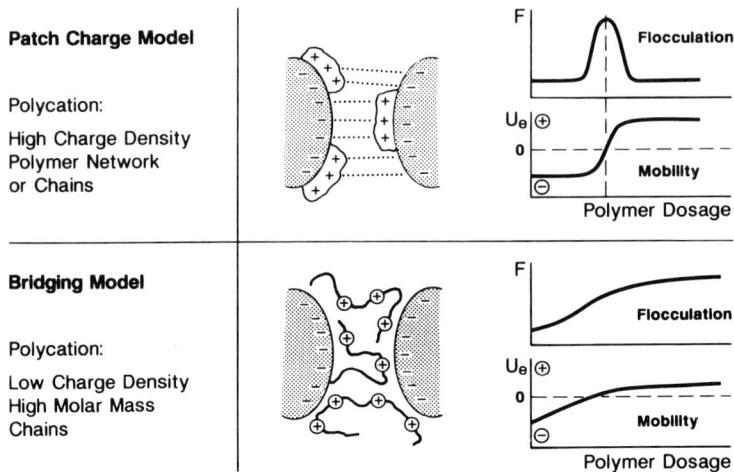

Figure 5.4   Electrokinetic effects and flocculation pattern in paper stock suspensions exposed to retention aids operating according to the patch charge and bridging mechanism (schematic).

PEI-type or low, PAM$^+$-type charge densities has shown that there are also characteristic differences in their electrokinetic effects and in the flocculation pattern induced at increasing concentrations [51, 62, 63, 66, 67, 68]. This is shown schematically in Figure 5.4. A simple explanation for these differences is provided by the fundamental differences in the structure of the adsorbates and the mechanisms by which polymers induce flocculation. The two principal flocculation mechanisms are termed bridging [51, 69, 70] and patch charge flocculation [13, 51, 71–73]. Retention aids with a low charge density and a chain structure act as bridging flocculants, especially if they have a high molar mass. The conformation of retention aids of this type in the adsorbed state allows them to string out, causing them to form numerous loops and tails. It takes a longer time for them to reconform, and the probability of particles being captured by the loops and tails is high.

Retention aids with a high positive charge density tend to be adsorbed in a planar structure with a large proportion of adsorbed trains [13]. They quickly reconform. At levels well below the saturation point, the interface at the substrate – which initially carries a uniform negative charge – is occupied by a mosaic of adsorbed, planar retention aid molecules which give rise to a bipolar interface consisting of positively and negatively charged patches [13, 71, 72]. Flocculation occurs as the result of attractive forces between oppositely charged patches of colliding particles. Maximum flocculation is normally provided if the substrate is occupied to around 50%. It would seem apparent that maximum flocculation depends on the isoelectric point being reached. It is important, however, to note that electrokinetic neutrality is not to be confused with electroneutrality in the sense of zero charge. The various sites at the

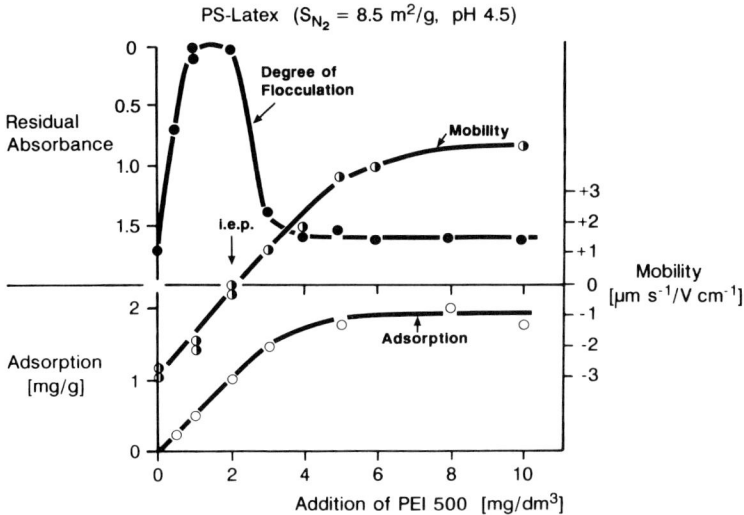

Figure 5.5 Electrokinetic effects and flocculation pattern of polystyrene latex ($d = 0.65\,\mu m$, $q = 2\,\mu C/cm^2$, $c = 2\,g/dm^3$) after adsorption of PEI 500 at $T = 25°C$. The degree of flocculation was determined by measuring the residual absorbance after sedimentation [13].

interface are positively or negatively charged, but in such a way that their electrokinetic action is cancelled out. Retention aids that function by means of mosaic formation differ from bridging flocculants in that adding too much retention aid leads to a redispersion of flocs, especially if polymer adsorption is assisted by shear-induced destruction of the floc structure [13, 33, 51].

A typical example of polymer-induced flocculation according to the patch-charge model is shown in Figure 5.5. It depicts the interactions between PEI and a polystyrene latex at pH 4.5 [13, 74]. Similar results are obtained for a modified PEI/bleached sulphite pulp system as shown in Figure 5.6 [75].

If, however, the charge density of the PEI retention aid is reduced by increasing the pH from 4.5 to 7.0 [13], it begins to show behaviour typically associated with bridging flocculants [75]. This emphasises the fact that, irrespective of their chemical nature, charge density plays a crucial part in determining the flocculation mechanisms of retention aids.

Because flocs are formed by different mechanisms, their structures vary. Patch-charge flocs are generally compact and their dimensions are considerably smaller than those of bridge flocs, which usually have a loose, voluminous structure. This affects retention and has important consequences for formation and drainage.

Another difference is highlighted by flocs' resistance to shear. Shear-induced deflocculation is reversible in the patch-charge mechanism [13], but bridging flocculants may be changed irreversibly by chain breakage preventing reflocculation, since the anchoring sites for the polymers at the interfaces remain occupied [13, 42].

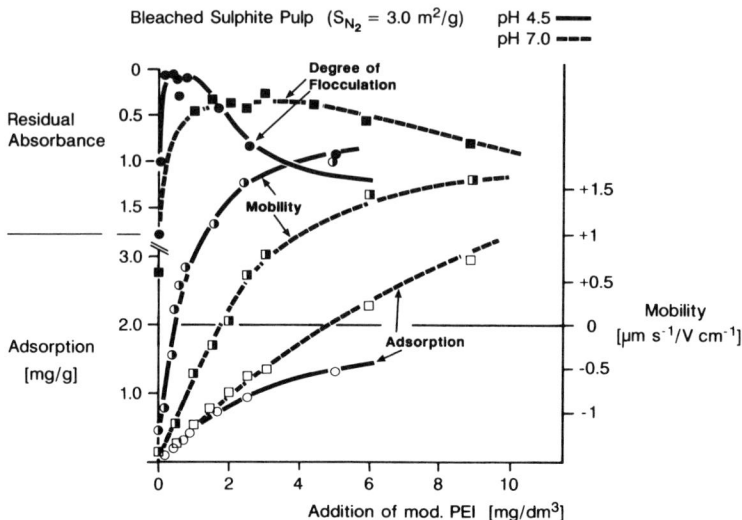

Figure 5.6 Electrokinetic effects and flocculation pattern of bleached sulphite pulp (BET specific surface $S_{N_2} = 3.0 \, m^2/g$, carboxylic content 0.9 mVal/100 g, c = 2 g/dm³, T = 25°C) after adsorption of modified PEI at pH 4.5 and pH 7.0, respectively [13, 51, 75].

Because of the kinetic effects described above, the changes in conformation of the molecules in the adsorbate layer and the varying degrees to which flocs are resistant to shear, the point at which retention aids are metered into the stock suspension plays an important role, and careful consideration needs to be given to this aspect when selecting retention aids [76].

Apart from the mechanisms described above, three other models have been put forward in the literature to describe the mode of action of retention aids. These can be essentially considered as being special cases of the two main models.

'Charge neutralisation flocculation' [77] is basically a special case of the patch-charge model. It can be assumed that the electrical charge at the interface will be cancelled out by ion exchange with polycations in stoichiometric proportions, and flocculation comes about by means of Van der Waals forces between uncharged particles. This model is plausible in situations in which a polycationic retention aid forms a nonionic polyelectrolyte complex with a polyanionic species. It is a prerequisite, however, that the structures of both reactants are potentially flexible, as is the case when cationic retention aids interact with disruptive anionic substances. Otherwise, cationic retention aids predominantly behave according to the bridging or patch-charge model when they are adsorbed on a solid surface with a rigid structure.

'Complex flocculation' [78–81] and 'network flocculation' [82, 83] may be considered as two special cases of the bridging model. These two special cases can be used to describe the behaviour of combinations of two or more polymeric retention aids. Anionic polyelectrolytes are able to cause disperse systems to flocculate if the latter are treated in advance with a cationic retention aid. The various retention aids combine in such a way that the cationic components adsorb onto the negatively charged interfaces establishing anchoring sites for the polyanions which now act as bridging flocculants. Oligomeric, hydrated aluminium salts can also be used as the cationic component.

Network flocculation has been discussed in conjunction with combinations of polyethylene oxides and phenolic resins. It is assumed that the two components form a three-dimensional network structure by means of hydrogen bonding. The suspended particles in the disperse phase are caught like fish in nets.

Most retention aids are polycationic, but the last two examples concern anionic and non-ionic substances. Retention comes about as the result of their adsorptive interaction with polyvalent metal ions or polycations, by the formation of hydrogen bonds or by unspecific interaction by means of Van der Waals forces. Of course, combinations of all three types are possible, with the main emphasis varying from process to process. Because of the variety of factors that influence flocculation and the changing composition of the stock, the choice of retention aids depends on empirical observations. Paper quality, especially formation, is an equally important point to be considered alongside retention and drainage.

5.4 Concluding remarks

The basic chemical and physical principles of retention provide a very useful guide to selecting the best retention aid from the many different types available.

Experiments to quantify the effects of pitch and other contaminants [48, 75], to measure adsorption and the associated electrokinetic factors, and especially for determining the kinetics of flocculation are very valuable tools. Measuring flocculation at different levels of shear turns out to be particularly useful [42–44, 84], given that the correlation between flocculation and retention has been established, as shown in Figure 5.7 [85].

New developments in measuring techniques employing fibre-optic sensors have made it possible for flocculation to be measured just as well *in situ* as in the laboratory [44, 84, 85]. This subject is discussed more fully in Chapter 12. Most recently a new laser-optical method for counting colloidally dispersed pitch has been introduced [86]. It enables a quick and reliable evaluation of the effectiveness of process chemicals used to control pitch, especially at high levels of closure at which paper machines are being operated.

It is to be expected that the planned use of the experimental methods

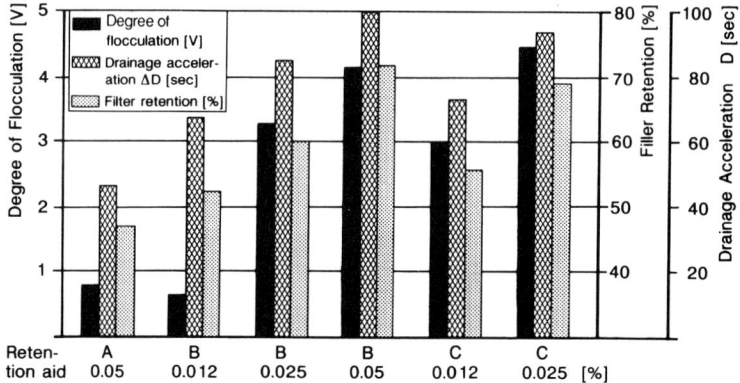

Figure 5.7 Interdependence of retention, drainage and flocculation of a paper stock (roto-gravure, pH 4.5), using retention aids of different nature (A, B, C) [85].

described above, combined with the substantial body of knowledge that has already been built up about the mode of action of retention aids, will enable significant progress in optimising the selection and application of retention aids. Economic and ecological benefits will play an equal part in these considerations.

References

1. Arnson, Th.R., *Tappi*, **65** (3) (1982), 125.
2. Strazdins, E., *Tappi*, **69** (4) (1986), 111.
3. Bottero, J.Y. and Fiessinger, F., *Nordic Pulp and Paper Res. J.*, 2–1989, **4** (1989), 81.
4. Batelson, P.G., Johansson, W.E., Larsson, H.M. and Svending, P.J., Canadian Patent 1154564 (1981).
5. Pye, D.J., US Patent 3052595 (1955).
6. Langley, J.G. and Litchfield, E., European Patent 17353 (1980).
7. Rutenberg, M.W., in *Handbook of Water-Soluble Gums and Resins*, ed. R.I. Davidson, McGraw-Hill Book Company, New York (1980), Chapter 22.
8. Reynolds, J.F. and Ryan, R.F., *Tappi*, **40** (1) (1957), 918.
9. Swift, A.M., *Tappi*, **40** (9) (1957), 224A.
10. Bikales, N.M., in *Water-Soluble Polymers*, ed. N.M. Bikales, Plenum Press, New York (1973), 213.
11. Volk, H. and Friedrich, R.E., in *Handbook of Water-Soluble Gums and Resins*, ed. R.L. Davidson, McGraw-Hill Book Company, New York (1980), Chapter 16.
12. Einhorn, A., *Liebigs Ann. Chem.*, **343** (1905), 207.
13. Horn, D., in *Polymeric Amines and Ammonium Salts*, ed. E. J. Goethals, Pergamon Press, Oxford (1980), 333.
14. Wilfinger, H., *Das Papier*, **2** (1948), 265.
15. Wilfinger, H. and Authorn, W., *Wochenblatt für Papierfabrikation*, **96** (1968), 201.
16. Poschmann, F.J., *Pulp and Paper Magazine, Canada*, **69** (1968), T 210.
17. Dick, C.R. and Ham, G.E., *J. Macromol. Sci.-Chem.*, **A4** (6) (1970), 1301.
18. Suen, T.J., Senior, A., Swanson, D.L. and Jen, Y., *J. Polym. Sci.*, **45** (1960), 289.
19. Espy, H.H. and Putnam, S.T., US Patent 3951921 (1976).

20. Butler, G., in *Polymeric Amines and Ammonium Salts*, ed. E.J. Goethals, Pergamon Press, Oxford (1980), 125.
21. Ottenbrite, R.M. and Shillady, D.D., in *Polymeric Amines and Ammonium Salts*, ed. E.J. Goethals, Pergamon Press, Oxford (1980), 143.
22. Hoover, F., *J. Macromol. Sci.-Chem.*, **A4** (6) (1970), 132.
23. Manley, J.A., US Patent 3141815 (1964).
24. Leung, P.S. and Goddard, E.D., *Tappi*, **70** (6) (1987), 115.
25. Braun, D.D., in *Handbook of Water-Soluble Gums and Resins*, ed. R.L. Davidson, McGraw-Hill Book Company, New York (1980), Chapter 19.
26. Eisenberg, H., *Biological Macromolecules and Polyelectrolytes in Solution*, Clarendon Press, Oxford (1976).
27. Kato, T., Tokuya, T., Nozaki, T. and Takahashi, A., *Polymer*, **25** (1984), 218.
28. Budd, P., *British Polymer J.*, **20** (1988), 33.
29. Pecora, R., ed. *Dynamic Light Scattering*, Plenum Press, New York (1985).
30. Wang, L. and Yu, H., *Macromolecules*, **21** (1988), 3498.
31. Lindquist, G.M. and Stratton, R.A., *J. Colloid Interface Sci.*, **55** (1976), 45.
32. Terayama, H., *J. Polym. Sci.*, **8** (1952), 243.
33. Horn, D., *Progr. Colloid & Polymer Sci.*, **65** (1978), 251.
34. Ebel, S. and Parzefall, W., *Experimentelle Einführung in die Potentiometrie*, Verlag Chemie, Weinheim (1975).
35. Bloys van Treslong, C.J. and Staverman, A.J., *Recl. Trav. Chim. Pays-Bas*, **93** (1974), 171.
36. Rice, S.A. and Nagasawa, M., *Polyelectrolyte Solutions*, Academic Press, New York (1961).
37. Bailey, P.L., *Analysis with Ion-Selective Electrodes*, Heyden, London (1978).
38. Rinaudo, M., in *Polyelectrolytes*, eds E. Selegny, M. Mandel and U.P. Strauss, D. Reidel Publishing Company, Dordrecht (1974), 157.
39. Shaw, D.J., *Electrophoresis*, Academic Press, London (1969).
40. Cohen Stuart, M.A., van den Boomgaard, Th., Zourab Sh.M. and Lyklema, J., *Colloids and Surfaces*, **9** (1984), 163.
41. Robertson, A.A. and Mason, S.G., *Pulp and Paper Magazine, Canada*, **55** (1954), 263.
42. Ditter, W., Eisenlauer J. and Horn, D., in *The Effect of Polymers on Dispersion Properties*, ed. Th.F. Tadros, Academic Press, London (1982), 323.
43. Wagberg, L., *Svensk Papperstidn.*, **88** (1985), R 48.
44. Eisenlauer, J., Horn, D. Linhart F. and Hemel, R., *Nordic Pulp and Paper Res. J.*, 4–1987, **2** (1987), 132.
45. Michaels, A.S., Mir, L. and Schneider, N.S., *J. Phys. Chem.*, **69** (1965), 1447.
46. Sandell, L.S. and Luner, P., *J. Applied Polym. Sci.*, **18** (1974), 2075.
47. Philipp, B., *Zellstoff und Papier*, **25** (1976), 102.
48. Palonen, H. and Stenius, P., *Proceedings EUCEPA Symposium, Warsaw* (1978) **90**; C.A. 91 (1979), 212836 f.
49. Cohen Stuart, M.A., Cosgrove, T. and Vincent, B., *Adv. Colloid Interface Sci.*, **24** (1986) 143.
50. Lindström, T. and Söremark, C., *J. Colloid Interface Sci.*, **55** (1976), 305.
51. Horn, D. and Melzer, J., in *Fibre-Water Interactions in Papermaking. Transactions of the Sixth Fundamental Research Symposium*, BPBIF, Oxford (1977), 135.
52. Kindler, W.A. and Swanson, J.W., *J. Polym. Sci.*, **A29** (1971), 853.
53. Alince, B. and Robertson, A.A., *J. Appl. Polym. Sci.*, **14** (1970), 2581.
54. Alince, B., *Cellulose Chem. Techn.*, **8** (1974), 573.
55. Lindström, T. und Wagberg, L., *Tappi*, **66**(6) (1983), 83.
56. Tanaka, H., Tachiki, K. and Sumitomo, M., *Tappi*, **62** (1) (1979), 41.
57. Evers, O.A., Fleer, G.J., Scheutjens, J.M.H.M. and Lyklema, J., *J. Colloid Interface Sci.*, **111** (1986), 446.
58. Wagberg, L. and Ödberg, L., *Nordic Pulp and Paper Res. J.*, 2–1989, **4** (1989), 61.
59. Horn, D. and Melzer, J., *Das Papier*, **29** (12) (1975), 534.
60. Onabe, F., *J. Appl. Polymer Sci.*, **23** (1979), 2909.
61. Melzer, J., *Das Papier*, **26** (7) (1972), 305.
62. Lindström, T., Söremark, Ch., Heinegard, Ch. and Martin-Löf, S., *Tappi*, **57** (12) (1974), 94.
63. Strazdins, E., *Tappi*, **57** (12) (1974), 76.
64. Hunter, R.J., *Foundations of Colloid Science*, Clarendon Press, Oxford (1987), 551.
65. Onabe, F., *J. Appl. Polymer Sci.*, **23** (1979), 2999.

66. Strazdins, E., *Tappi*, **55** (12) (1972), 1691.
67. Moore, E.E., *Tappi*, **58** (1) (1975), 99.
68. Anderson, R. and Penniman, J.G., *Paper Trade J.*, **23** (September 1974), 22.
69. Healy, T.W. and LaMer, V.K., *J. Phys. Chem.*, **66** (1962), 1835.
70. Pelton, R.H. and Allen, L.H., *Colloid Polym. Sci.*, **261** (1983), 485.
71. Kasper, D.R., Ph.D. Thesis. California Institute of Technology (1971).
72. Gregory, J., *J. Colloid Interface Sci.*, **42** (1973), 448.
73. Goossens, J.W.S. and Luner, P., *Tappi*, **59** (2) (1976), 89.
74. Eisenlauer, J. and Horn, D., *DVGW-Schriftenreihe*. Wasser Nr. 42. Eschborn (1985), 59.
75. Horn, D., *Zellstoff und Papier*, **28** (1979), 129.
76. Van de Ven, T.G.M. and Mason, S.G., *Tappi*, **64** (9) (1981), 171.
77. Vincent, B., *Adv. Colloid Interface Sci.*, **4** (1974), 193.
78. Vincent, B. and Whittington, S.G., in *Surface and Colloid Science*, ed. E. Matijevic, Plenum Press, New York, **12** (1982), 1.
79. Ries, H.E. and Meyers, B.L., *J. Appl. Polym. Sci.*, **15** (1971), 2023.
80. Britt, K.W., *Tappi*, **56** (3) (1973), 83.
81. Moore, E., *Tappi*, **59** (6) (1976), 120.
82. Lindström, T. and Glad-Nordmark, G., *J. Colloid Interface Sci.*, **97** (1984), 62.
83. Pelton, R.H., Allen, L.H. and Nugent, H.M., *Tappi*, **64** (11) (1981), 89.
84. Eisenlauer, J. and Horn, D., *Colloids and Surfaces*, **14** (1985), 121.
85. Linhart, F., Horn, D., Eisenlauer, J. and Hemel, R., *Wochenblatt f. Papierfabrikation*, **115**–**8** (1987), 331.
86. Kröhl, T., Lorencak, P., Gierulski, A., Eipel, H. and Horn, D., *Nordic Pulp and Paper Res. J.*, 1–1994, **9** (1994), 26.

6 Dry-strength additives

J. MARTON

6.1 Introduction

Dry strength is an inherent structural property of a paper sheet which is due primarily to the development of fibre to fibre bonds during consolidation and drying of the fibre network. Paper strength is dependent on the strength of individual fibres, the strength of interfibre bonds, the number of bonds (bonded area) and the distribution of fibres and bonds (formation). Fibre to fibre bonds are usually weaker than the strength of individual fibres until the latter becomes the limiting factor in a well-bonded sheet [1]. Paper strength additives may bring about improvement in one or more of the above factors, although it may be assumed that they are unlikely to affect the strength of single fibres.

Mechanical entanglement leading to fibre flocculation is a first step in a long chain of events forming the stable network of fibres [2]. One consequence of beating is the increased fibrillation of the fibre surface, enhancing the probability of entanglement and producing a great number of microfibrils capable of actively contributing to the formation of additional fibre–fibre bonds. Various forces may participate in fibre to fibre bond formation. The most important being that of hydrogen-bond formation, although other bonding forces such as covalent, ionic (electrostatic) and van der Waals forces also may be operative.

Hydrogen bonds are water sensitive. Although rather weak bonds, they may still act in concert and can effect significant inherent paper strength. They are active only over short distances and, hence, the bonding surfaces must be in molecularly close contact.

There are several ways, in papermaking practice, to increase sheet strength. This may be by furnish modification, for example the inclusion of a higher proportion of long fibres, a reduction in the amount of inert filler, or by the application of strength additives. It may also be achieved by process modification, for example the use of additives, alkaline pH, increased wet pressing, or the application of starch at the size press or in the coater. Refining (beating) is, however, perhaps the most frequently used tool in papermaking for increasing tensile and related strength properties. It may not always be appropriate, however, for in addition to the cost of the electrical energy consumed, refining usually slows down drainage on the wire, increases the density of the resulting sheet and decreases its porosity. With the loss of bulk, stiffness may also have been impaired, and tearing resistance decreases. The increased bonding also reduces opacity.

Certain chemical additives, when introduced into the furnish prior to paper formation, constitute an alternative method to beating for generating paper strength, whilst maintaining other important property combinations. Dry-strength additives are usually water soluble, hydrophilic natural and synthetic polymers. The commercially most important are:

- cationic starch
- natural vegetable gums
- acrylamide polymers.

There are also others which are used in specialised applications. These are:

- unmodified and anionically modified starches
- soluble cellulose derivatives such as carboxymethyl cellulose
- dual purpose wet- and dry-strength resins
- polyvinyl alcohol, latex and other polymers.

These additives may also have other effects. Addition, for instance, of only 0.01–0.02% polyacrylamide polymer or of about 0.5% cationic starch may greatly improve fines retention and, in some cases, also drainage on the wire. Increased levels of addition of these and other wet-end additives, for example 0.5% acrylic polymer or 1–2% cationic starch, may cause substantial improvement in sheet strength, but it may also adversely affect formation. The superior bonding properties of natural gums, cationic starches and synthetic polymers can be employed for various purposes [3]:

(a) Improvement of tensile, burst and fold.
(b) Improving tear resistance by reducing refining whilst maintaining tensile at the desired level.
(c) Utilising higher percentages of lower strength pulps and pigments.
(d) Increasing production rate by speeding up refining and drainage.
(e) Attainment of better fibre and fines distribution and/or sheet formation (these tend not to be strongly cationic additives which may impair formation).
(f) Attainment of certain desirable combinations of properties which are difficult to obtain otherwise. For instance, high strength combined with high porosity, improved hygroexpansivity and reduced loss in optical properties.

Recent trends in papermaking affect paper strength in numerous ways. Table 6.1 summarises and complements some factors already mentioned.

Table 6.1 Effect of some papermaking factors on strength

Trend	Effect on strength
Shift in wet-end pH toward neutral/alkaline	+ +
Increased recycled fibre usage	− −
Replacement of filler clay with calcium carbonate	+
Increased filler load	−
Increased cationic starch usage	+
Increased contaminants from high yield pulps and recycled fibres	−
Increased anionic interference from system closure	−

The attention of this chapter will be focused on selected aspects of the application and mechanism of action of the three most generally used dry-strength additives: cationic starch, vegetable gums and polyacrylamide based polymers. More detailed information and comprehensive literature reviews are available elsewhere [3–6].

6.2 Cationic starch

6.2.1 *Use of the starch adhesive*

Starch is one of the oldest and still most commonly used adhesives today applied for strengthening fibre bonding in paper. Wet-end retention of the native unmodified jet-cooked (anionic) starch is quite low ($< 40\%$) on modern fast machines. In order to increase its retention, great efforts were made in the 1960s to use natural pearl starch in a physical blend with cationic agents or to use it together with cationic starch as a less expensive extender.

Experience [7] indicated that the cationic components of the mixture rapidly separated and preferentially adsorbed on to the furnish, thus effectively enriching the non-adsorbing pearl starch content of the recirculated white water. Only permanent chemical combination of the cationic reagent with starch could make its use practical for modern papermaking.

6.2.2 *Cationic starch preparations*

There are today many commercial preparations available for the paper maker to choose from [5,8,9]. The molecular weights of the unmodified starches may range from hundreds of thousands to as high as a hundred million daltons (for high amylopectin waxy maize starch). The most frequently converted starches originate from (in the order of increasing molecular weight): yellow corn < tapioca $\leqslant$ potato $\ll$ waxy maize (high amylopectin) corn starch.

The cationic starches are most effective if their polyelectrolyte character is enhanced with the proper balance of anionic and cationic charges. Phosphate (phosphonic acid) groups are the customary source of these anionic charges. Since corn starch has only a low native phosphate content, it is customary to insert into it the phosphoric acid residue separately (amphoteric starches).

The cationic charge can be introduced by reactive monomeric or polymeric tertiary amine or quaternary ammonium derivatives. Tertiary amines require protonation to become charged and are cationic only below pH 7. The converted starches usually contain 0.1–0.4% nitrogen, their cationic charge being roughly proportional to the nitrogen content. The starches can be converted in the uncooked state (by the supplier) or in soluble form during the cooking process (at the mill).

The most commonly used tertiary reagent is diethylamine ethyl chloride, hydrochloride.

$$(C_2H_5)_2N.CH_2CH_2Cl.HCl$$

The most frequently used quaternary reagent is epoxy propyl trimethyl ammonium chloride.

$$CH_2\text{–}CH\text{–}CH_2\text{–}\overset{+}{N}(CH_3)_3 \cdot Cl^-$$

A new type of cationic wet-end starch has recently been commercialised as a paper strength agent. It has blocked aldehyde groups (hemiacetals) from which the aldehyde groups are freed just before introduction to the papermaking system [10]. The nascent aldehyde is able to interact by crosslinking with surface -OH groups of cellulosic fibres. Unlike conventional cationic starch, the aldehyde-starch also temporarily increases wet strength in both dried-and-rewetted and never-dried paper, and this may be a desirable asset in certain applications.

6.2.3   Mechanism of strength development

Moeller [11] proposed that the adsorption of cationic starch creates new bonding sites on the fibre surface which are stronger than the original fibre to fibre bonds. In other words, the strength increase is due to additional fibre to fibre bonds and not to the strengthening of existing bonds. Fibres may adsorb 4–5% cationic starch, the first 1–2% of which would be retained on the most active areas of the fibre surface, and thus the most likely potential bonding sites. This irreversible retention occurs at the initial contact, and final strength development, unlike with gum or polyacrylamide, is not dependent on migration of starch to the fibre crossings during drying. In the recent work of Howard and Jowsey [12], the initial 0.5% had the greatest effect. Cationic starch was found to increase tensile strength of paper by increasing the bond strength per unit optically bonded area, rather than relative bonded area.

Fibre to fibre bonding is usually explained by the formation of hydrogen bonds during drying [13]. Hydrogen bonds are only effective over a very short distance, approximately 0.3 nm. Rough fibre surfaces have asperities larger than that, thus physically preventing the formation of hydrogen bonds. Addition of 1–2% cationic starch may fill out these asperities with an adhesive matrix, thereby creating new bonding areas as indicated earlier. This is particularly true of filled papers, where the applied pigment is competitively adsorbed by the fibre surface. Potential fibre to fibre bonding sites are blocked and hydrogen bond formation between fibres is prevented.

The swelling capacity and thus the H-bonding ability of the recycled fibres is diminished. The drying phase in the first papermaking cycle increases stiffness, and reduces fibre flexibility and interfibre bonding capability [14]. This effect is more pronounced in chemical pulps than in mechanical pulps. Another negative aspect of recycled fibres is the increased level of fines from the refining of these more brittle fibres. These undesirable consequences may be partially ameliorated by increasing the addition level of cationic starch.

6.2.4 Highly filled papers

It is common practice in fine paper manufacture to increase, for various reasons, the filler content of the sheet and to counteract strength loss by addition of wet-end starch or other dry-strength agents. In certain commercial applications, the cationic starch is preadsorbed onto the clay filler in order to enhance the strengthening effect (HILOAD system). Lindström [15] proposed that massive strength improvement can be achieved and/or a high filler load can be tolerated by the use of excessive amounts of cationic starch. Such additions are well above the adsorbing capacity of the furnish, and they achieve their effect by the unadsorbed starch being precipitated onto the fibre surface by addition of an anionic polyacrylamide. High filler loads, without a compensating high starch content, would naturally reduce paper strength. The high filler content also counteracts opacity loss due to the high starch content. Lindström reports that acceptable sheet properties were achieved in super-filled paper structures which contained up to 50% filler loading. Significant increases in wet-end clay filler addition in paper mills have not yet proved to be practical. However, elevated levels of calcium carbonate fillers in ground and precipitated forms are becoming standard in fine papers. The inevitable strength loss which this causes is easily compensated by increasing the addition level of cationic starch.

6.2.5 Adsorption of starch

Efficient adsorption of starch on furnish components is imperative in order to achieve economically the expected strength advantages and to lessen waste disposal problems caused by unretained starch. Adsorption of most anionic or non-ionic starches on virgin fibre surfaces is a fairly reversible process. Whilst retrogradation may reduce the solubility of certain high amylose starches, usually over 80% of the adsorbed starch can be removed simply by heating the aqueous suspension. When cationic starch is adsorbed onto the fibre, its adsorption is virtually irreversible. Over 85% remained on the fibre in hot water, and it could be redissolved only by a strong acid wash [16].

The two major constituents of cationic starch, namely cationised amylose and amylopectin, exhibit different adsorption affinity towards cellulose [17]. At low electrolyte concentration, cationic amylose adsorbs preferentially due probably to its more flexible linear molecular structure. The adsorption of both components strongly decreases with increasing electrolyte and increasing cellulose concentrations, indicative of the importance of charge interactions in the adsorption process.

Retention of certain cationic polymers on cellulose has been explained by ionic interaction between the cationic additive and cellulose-COOH groups [3,4]. A similar interaction has been demonstrated between cellulose and cationic starch [16]. In this experiment, alpha-cellulose was carboxymethylated to different degrees and the number of carboxyl groups was determined by the standard methylene blue adsorption method (Tappi T237). Titration of these carboxymethylated cellulose preparations with a low molecular weight cationic surfactant (molecular weight 318; C^+16) showed a stoichiometric correspondence between the surfactant and the reactive –COOH groups in the CMC-cellulose preparations, thus confirming that all –COOH groups were available for low molecular weight reactants (Figure 6.1). Adsorption studies with cationic potato starch showed, however, that whilst the increasing

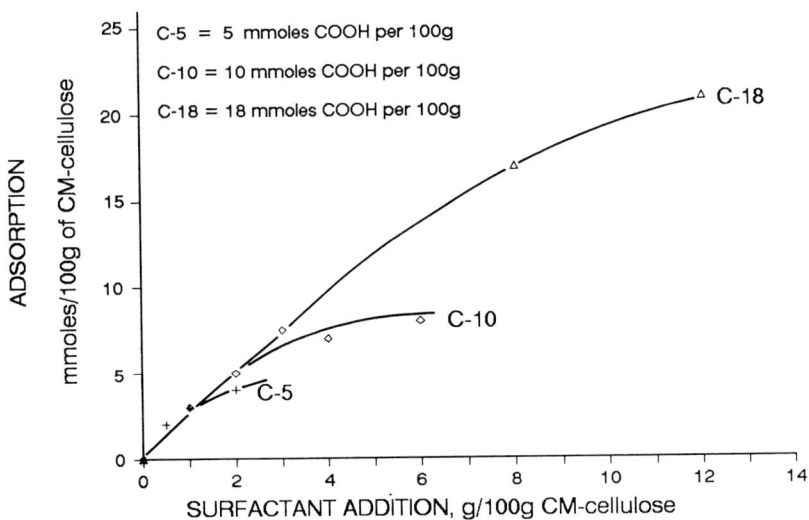

Figure 6.1 Adsorption of a cationic surfactant ($C_{16}H_{31}N^+$ (CH_3)$_3$) by CM-cellulose after 30 minutes stirring at pH 5.

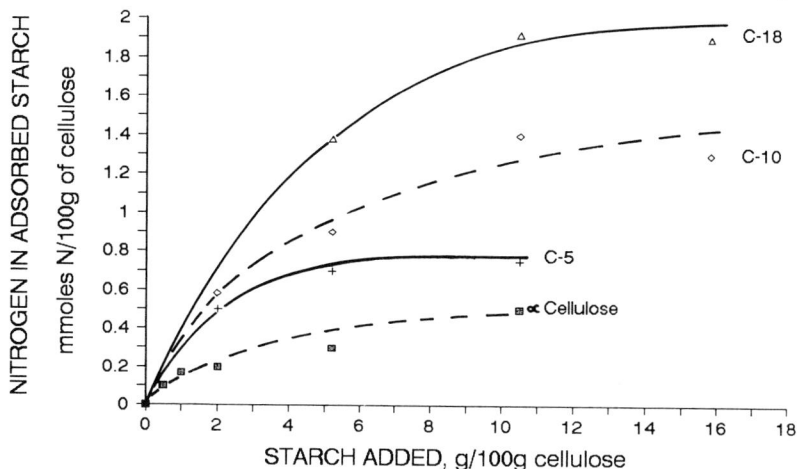

Figure 6.2 Adsorption of cationic potato starch (N = 0.35%) by CM-celluloses after 30 minutes stirring [16]. (Note: C-5, C-10, C-18 as in Figure 6.1).

−COOH content enhanced starch adsorption, only about 10% of the fibre-COOH groups had participated in the starch-fibre interaction (Figure 6.2).

The high molecular weight starch could not penetrate the small capillaries of the swollen CMC-cellulose preparations and, consequently, only a portion of the total surface area, and therefore total −COOH groups, was available for starch adsorption.

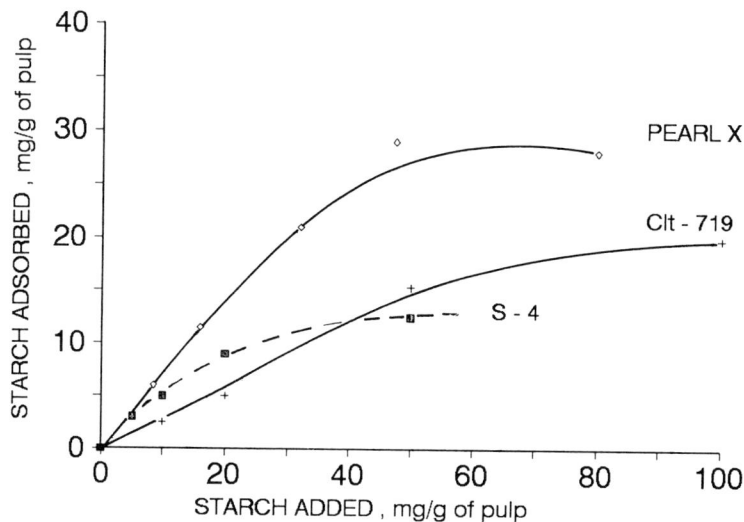

Figure 6.3 Adsorption of starches by cationised (0.3% N) pulp [16]. Note: Pearl X = regular anionic pearl starch; Clt-719 = ethoxylated corn starch; S-4 = cationic potato starch).

In a related study [16], alpha-cellulose was reacted with trimethyl ammonium epoxy propane in the presence of alkali (15 mmoles per 100 g fibre). Figure 6.3 shows that cationic starch (having a similar, 0.3 N content) adsorbed onto the cationised fibre to a lesser extent than on normal anionic fibre. Its adsorption by cationic fibre was also less than the adsorption of slightly anionic pearl starch or strongly anionic starch ether preparation. These latter two starches would not be expected to be adsorbed significantly by normal anionic fibres and would be quite ineffective in the common paper-making furnish unless the fibres were made cationic in a preceding step.

With regard to the strong interaction between oppositely charged starch and fibres and to the slight interaction between similarly charged starches and fibre, it was concluded that both electrostatic (ionic) bonding and hydrogen bonding participate in the process of adsorption of starch on fibre, ionic bonding being usually the dominant force in the cationic starch and fibre interactions. If other cationic species, such as oleated aluminium cations, are present and can compete with cationic starch for the fibre adsorption sites, the net result is a reduction of starch adsorption on fibre or on other anionic furnish components [16].

6.2.6   *Adsorption of cationic starch on furnish components*

Cationic starch is usually applied in fine paper furnishes at a level of 0.25–2.50%. Excessive starch may reverse the charge on fillers and fines thereby reducing their retention. For both environmental and economic reasons, it is important that only small amounts of non-adsorbed starch remain in the white water. Figure 6.4 shows the adsorption isotherms of a cationic, medium charge density (0.35% N) potato starch on various commonly used furnish ingredients. The adsorption isotherms are of the Langmuir type [18]. Table 6.2 and Figure 6.4 show the saturation values (a_s) of the isotherms together with the hydrodynamic surface area values for specific furnish components. It

Table 6.2   Saturation values for cationic starch adsorption by common furnish components [18]

Adsorbent	Hydrodynamic surface area (m^2/g)	Saturation value (a_s, mg/g)
Fibre fraction	1.2	16.0
Pulp blend	2.1	42.0
Fines fraction	8.0	65.0
Filler clay	2.0	17.0
No. 2 coating clay	12.0	59.0
Premium coating clay	13.0	55.0
Ultragloss 90	22.0	64.0
Paperad	7.0	13.0
Rutile-TiO_2	10.0	17.0

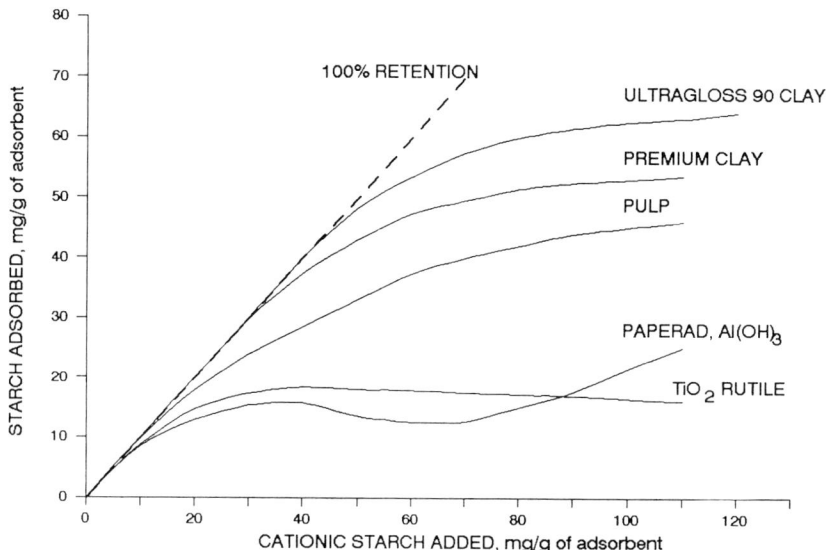

Figure 6.4 Adsorption isotherms for cationic starch by blend pulp and various fillers at pH 5[18]. (Note ULTRAGLOSS 90 and PAPERAD are trademarks of Engelhard Minerals and Chemicals Corporation, and REYNOLDS Metals Corporation respectively).

has been assumed that the hydrodynamic surface area values approximate best to the surface areas available for starch adsorption.

The saturation adsorption capacity (a_s) of the furnish is roughly proportional to the available surface areas, with some notable exceptions, suggesting that surface composition and charge of the interacting species also play a role. It is interesting to note that both the applied TiO_2 and alumina pigments are slightly cationic at the pH of the experiment (pH 5). These pigments also exhibit inhibited adsorption of the cationic starch in a manner comparable with cationised pulp (Figure 6.3). The irregular shape of the isotherm for the alumina pigment in Figure 6.4 can be rationalised by the fact that it was not a pure preparation but contained 20% (anionic) calcined clay.

6.2.7 Distribution of cationic starch in the furnish

The saturation values (a_s) shown in Figure 6.4 and Table 6.2 have been obtained at rather a high level of starch application. In practice, cationic starch is usually applied at much lower levels, where the adsorption efficiencies are close to 100% and the values mostly fall onto the initial steep part of the adsorption curves. Cationic starch, if captured by one adsorbent, is not readily desorbed and is thus not available to a second adsorbent.

In most mill situations, cationic starch is added to a blend of furnish components. In order to understand or predict the behaviour of the furnish, it

Table 6.3   Distribution of cationic potato starch (1.3%) in a model fine paper furnish at pH 5 [18]

Component	Composition	Adsorption intensity	Will adsorb (eff. 95%)		Expected starch concentration (%)
			Value	% Ratio	
Fibre	0.69	1	0.69	33	0.57
Pulp fines	0.15	5	0.75	37	3.0
Clay	0.16	4	0.64	30	2.2
Furnish	1.0	–	–	–	1.2

would be helpful to know the precise amounts adsorbed by each component at the practical levels of addition. Such determinations were carried out on binary components mixtures [18] and were also compared to the ratio of the a_s saturation values. It was concluded that the a_s values give a good estimate of the relative driving forces operative along the whole adsorption isotherm and an estimate can be made of the likely adsorption intensities on different adsorbing surfaces. Table 6.3 shows the computed distribution of starch in a model pigment-loaded furnish by use of such derived adsorption intensity values [18]. The table shows that the 70% fibre portion adsorbs only one third of the applied starch, the remainder being almost equally adsorbed onto the fines and filler in the furnish. The expected starch concentrations are therefore much higher on the filler and pulp fines, leading to uneven starch and charge distribution in the furnish. Large amounts of fillers and/or fines in a furnish cause a reduction in the starch present on the fibres, and hence reduce the amount available to strengthen interfibre bonding.

It follows that the order of mixing of the furnish components with starch, the addition sequences and the location of application points become potent tools for emphasising selected production goals. Table 6.4 indicates the results of a laboratory handsheet experiment where the addition sequence was varied [7]. Sequence B (i.e. adding starch to fibres in the thick stock before filler addition) offers the advantage of higher filler retention, and Sequence C (i.e. adding starch to fillers separately, before fibre addition, or to the thin stock) emphasises the improvement of sheet strength. Whilst starch is usually most effective as a bonding agent when it is located on long fibres (see below), in sequence, the cationic starch aggregates the filler and thus reduces its debonding effect at the price of some opacity loss. This is similar to the HILOAD technique discussed earlier.

In a recent related study, Stratton [19] investigated the dependence of sheet properties on the location of adsorbed polymers using polyamine epichlorohydrin and carboxymethyl cellulose. Strength aids added separately to classified pulp components produced the highest strength when adsorbed on the long fibre only. The optimum location of the strength

Table 6.4 Effects of addition sequence on retention and paper hand-sheet properties [7].

Property	Effectiveness of mixing method*	Property range
Starch retention	A < B < C	90–95%
Filler retention	C < A < B	71–82%
IGT pick resistance	A < B < C	15–23
Internal bond strength	B < A < C	200–240
Fold endurance	B < A < C	80–105
Burst strength	B < A < C	41–44
Opacity loss	A < B < C	0.5%

*Mixing method A: Starch added to a mixture of fibre + filler
 Mixing method B: Filler added to a mixture of fibre + starch
 Mixing method C: Fibre added to a mixture of filler + starch

aid has been on the long fibre, not on the fines. The much larger percentage of strength enhancement for classified compared to whole pulps was shown to be due to a shift in the mechanism of failure. Strength in untreated sheets is governed by the bond strength between the fibres. With the addition of a strength aid, the bond strength is increased and the individual fibre strength becomes the limiting factor. Wet-end addition of a strength aid was found to be more effective than external application with a size press. It is considered to be important to have the polymer within the crossover region between two fibres (i.e. the bonded area) and not just around the periphery of the bond. The main function of the strength aid is to increase the bond strength within the bonded area.

Another potentially applicable observation [20] is that high shear can promote deposition of charged high molecular weight polymers onto the long fibres. Such high shear conditions may exist in the fan pump or at the pressure screens. Whilst strength aids are usually applied in the back-system in rather low shear areas, high shear may also be created there within properly constructed (e.g. in medium consistency) application pumps [21].

As indicated, dry-strength agents are usually more efficient when deposited on to long fibres rather than on any other furnish components.

Proper control of starch addition rate and its directed distribution can therefore help to optimise either interfibre bonding or retention, making cationic starch a versatile and useful wet-end additive.

6.3 Vegetable gums

Vegetable gums are less frequently used than starch nowadays in the paper mill. Those most commonly used are guar and locust bean gums. The backbone chains in these gums are rigid, rod-like mannose units with short galactose

side-chains. Their molecular weights are of the order of 200 000–300 000
daltons. Karaya gum, a less frequently applied mucilage, has a somewhat more
complex side-chain structure, with a molecular weight close to 10 million [22].
The most important use of these gums is in unbleached kraft mills.

6.3.1 *Mechanisms*

The gums are very hydrophilic polysaccharides and their aqueous solutions
are quite viscous. The molecules are adsorbed through van der Waals forces
and H-bonding to fibre surfaces, and can, after drying, enhance molecular
contact between fibres. Leech [23] found that as little as 0.07% (based on fibre)
locust bean gum was an effective strengthening agent. He estimated that 60%
of the benefits obtained were attributable to improvements in bond strength,
25% to improved formation and 15% to increased numbers of bonds. It was
also postulated [4] that the capillary forces of drying (Campbell forces) move
the well-dispersed gum molecules towards the fibre cross-over areas where
they are most effective in strengthening bonding. The high surface area of the
fines attracts a major portion of the applied gums in a similar manner to the
distribution of cationic starch. Modified guar gums containing cationic
groups have been prepared so as to improve their adsorption. Such cationic
gums are also applicable in filled fine paper furnishes. Suppliers claim
synergistic action with other wet-end additives, thus improving both retention
and drainage [24].

   As formation aids, gums may act as colloidal protecting agents by reducing
the rate of flocculation of long-fibred pulps. They may also act as a lubricant
for the fibre surface and assist by giving a better distribution of anionic charges
along the fibre surface [25]. Mostly guar and locust beam gums have been
used to improve fibre bonding and formation. Guar gum also reduces the
friction of fibre suspensions in turbulent flow. As fibre dispersants, gums with
unbranched or slightly branched structures having $1 \rightarrow 4$ glycosidic linkages
are found to be the most effective. The most potent deflocculation agent is the
very high molecular weight deacetylated karaya gum when used in the
presence of alum [25].

6.4 Polyacrylamide resins

Polyacrylamides (PAM):

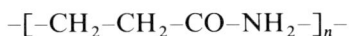

$$-[-CH_2-CH_2-CO-NH_2-]_n-$$

have been very versatile chemical aids for the paper industry since the mid
1950s. Depending on molecular weight and the nature and amount of charge-
bearing copolymers or substituents, they have found many and varied

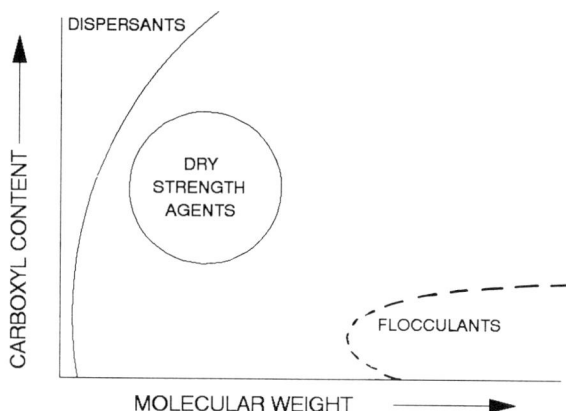

Figure 6.5 Effect of carboxyl content and molecular weight on the application of acrylamide-based polymers [26].

applications. Figure 6.5 schematically indicates the three most important areas of application as a function of their –COOH content and molecular weight [26]. These applications range from pigment and scale dispersion, through dry-strength agents to flocculation and retention. For strengthening effects, the polymer molecule must be long enough to ensure effective adsorption and to provide multiple sites for H-bonding without bridging and flocculation of the fibres [27]. The practical molecular weight range is between 100 000 and 500 000 daltons. Anionic (–COOH) or cationic substitution is typically between 5 and 10 mole % with respect to the amide group.

Unlike refining, no change in apparent density is observed when any of the PAMs are used for dry-strength development. In addition to increases in tensile, burst, fold and internal bond strength, stiffness and ring-crush are also improved. Due to their relatively high price, the use of PAMs can only be justified where their special advantages can be exploited.

Anionic PAMs are made fibre-substantive by using alum or a mediating cationic polymer. 0.25% anionic PAM was found to improve the tensile strength in a moderately beaten pulp by 18%, and 0.5% gave a 32% increase [28]. When alum is used at a pH of about 4.5–4.8, partially neutralised (30-50%) alumina flocs are most effective for achieving good adsorption to the fibres. For anionic PAM, the minimum alum requirement is 1.25%. In the case of clay-filled papers, double this amount is needed. It is common practice to add rosin size and alum before the refiners and use PAM afterwards. The higher the stock temperature, the closer the alum should be added to the headbox so as to preserve the integrity of the alumina floc. Where maximum drainage is required, the rosin should be added after the screens in order to avoid shear-induced degradation. Naturally, PAM strength additives are not as shear-sensitive as their higher molecular weight retention aid relatives.

The practical choice of PAM dry strength resin for each individual system will depend on such factors as electrokinetic charge, the amount of alum or other cationic components, and pH and temperature.

Cationic PAM resins were commercialised around 1967 and are widely used in various paper and board grades. Their main benefits are applicability over a wide pH range, no requirement for alum and adaptability to all kinds of furnishes. Cationic PAMs can also assist in drainage and can serve as retention aids.

Glyoxalated crosslinked PAM resin also is a highly effective newly developed dry-strength additive which also develops temporary wet strength in rewetted sheets, thus further increasing the choice of tools for the papermaker [29].

6.4.1 *Mechanisms*

Reynolds [5, 26] has proposed that the PAMs provide additional bonds between fibre surfaces where the distance is too great for H-bond formation between adjacent OH groups on cellulose. These new bonds are long and flexible enough to work without establishing optical contact between the bonding sites. The most convincing evidence for his hypothesis is the ability of PAMs to double strength with very little change in density, porosity or light scattering. Short-bridging from one surface to another also may explain why PAM is particularly effective in increasing ply-bond strength.

Further information on these and other dry-strength additives is available in various monographs and the reader is referred in particular to [5] and [6].

References

1. Page, D.H., *Tappi*, **52** (4) (1969), 674.
2. Kerekes, R.J., *J. Pulp Paper Sci.*, **9** (3) (1983), TR 86.
3. Swanson, J.W., *Tappi*, **44** (1) (1961), 142A.
4. Schwalbe, H.C., in *Consolidation of Paper Web, Cambridge Symposium*, ed. F. Bolam, London: Technical Section BPBMA (1966), 345.
5. *Dry Strength Additives*, ed. W.F. Reynolds, TAPPI Press, Atlanta, GA (1980).
6. *Pulp and Paper, Chemistry and Technology*, 3rd edn, ed. J.P. Casey, Vol. III, J. Wiley & Sons, New York (1981).
7. Marton, J., unpublished data.
8. Maher, S.L., in *TAPPI 1985 Adv. Topics in Wet End Chem. Sem.*, TAPPI Notes, Atlanta, GA (1985), 21.
9. Harvey, R.D., in *TAPPI 1987 Adv. Topics in Wet End Chem. Sem.* TAPPI Notes, Atlanta, GA (1987), 83.
10. Solarek *et al.*, US Patent 4675394 (June 1987).
11. Moeller, H.W., *Tappi*, **49** (5) (1966), 211.
12. Howard, R.C. and Jowsey, C.J., In *Proc. Intern. Paper Phys. Conf.* Montreal, Quebec (1987), 217.
13. Nissan, A.H., in *Lectures on Fiber Science*, ed. W.C. Walker, CPPA-TAPPI Joint Textbook Comm., Montreal, Quebec (1977), 31.
14. Minor, L. and Atalla, R.H., in *Proc. Mat. Res. Soc. Symp.*, **266** (1992), 215.

15. Lindström, T. and Floren, T., *Svensk Papperstdn.*, **87** (12) (1984), R 99.
16. Marton, J. and Marton, T., *Tappi*, **59** (12) (1976), 121.
17. van Steeg, H.G.M. *et al.*, *Nordic P&P Res. J.*, **4** (2) (1989), 173.
18. Marton, J., *Tappi*, **63** (4) (1980), 87.
19. Stratton, R.A., *Nordic P&P Res. J.*, **4** (2) (1989), 104.
20. van de Ven, T.G.M., in *Fundamentals of Papermaking*, eds C.F. Baker and V.W. Punton, *9th Fundamental Res. Symp.* Cambridge (1989), MEP, London (1989), 471.
21. Marton, J., *Proc. EUCEPA Symp. on Additives, Pigments and Fillers*, Barcelona, Spain (Oct. 1990), 105.
22. Werdouschegg, F.M.K., in *Dry Strength Additives*, ed. W.F. Reynolds, TAPPI Press, Atlanta, GA (1980), 67.
23. Leech, H.J., *Tappi*, **37** (8) (1954), 343.
24. Coco, E., in *TAPPI 1980 Retention and Drainage Sem.*, TAPPI Notes, Atlanta, GA (1980).
25. Hofreiter, B.T., in *Pulp and Paper, Chemistry and Technology, 3rd edn*, ed. J.P. Casey, Vol. III, J. Wiley & Sons, New York (1981), Chapter 14, 1475.
26. Reynolds, W.F., *Tappi*, **44** (2) (1961), 177 A.
27. Reynolds, W.F. and Wasser, R.B., in *Pulp and Paper, Chemistry and Technology, 3rd edn*, ed. J.P. Casey, Vol. III, J. Wiley & Sons, New York (1981), Chapter 13, 1448.
28. Reynolds, W.F., in *Dry Strength Additives*, ed. W.F. Reynolds, TAPPI Press, Atlanta, GA (1980), Chapter 6, 125.
29. Farley, C.E., *TAPPI Adv. Topics in Wet End Chem. Seminar*, Memphis, TN (1987). p. 91.

7 Wet-strength chemistry

N. DUNLOP-JONES

7.1 Introduction

Paper is a layered mat consisting of a network of cellulose fibres bonded together. Each of the fibre–fibre contacts is held together by intermolecular forces (van der Waals; hydrogen bonding) which are very sensitive to water, the extent of bonding decreasing steadily as the water content of the paper increases [1]. The water wets the fibres which swell; the bonds are broken leaving somewhere between 3% and 10% of the original dry strength (at 50% relative humidity). It is thought that some of this residual strength results from covalent fibre–fibre bonds [2,3]. Apart from the influence of the physical properties of the fibre network, the strength of the paper matrix is dependent on the bonded area, and this can be increased by forcing the surfaces together as in wet pressing, or by increasing fibre flexibility and swelling. Other solvents also affect strength, and their effects are essentially related to their ability to swell cellulose [4–6].

As there is a need for paper products which retain some strength when subjected to high humidity or when soaked in water, chemical wet-strength resins have been developed. It is now possible to produce paper products with wet strengths of over 50% of the dry strength, although 20–40% is more typical. In practice, paper products with a wide range of dry strengths are found because of their different weights, densities and degrees of interfibre bonding. Therefore it is impossible to define wet strength in absolute terms and it is more usual to express the wet-to-dry tensile as a percentage. It has been suggested [7] that papers with wet-tensile strengths of more than 15% of the dry tensile should be considered wet-strength papers.

It is also possible to further subdivide wet-strength papers based on the permanence of the strength. Untreated paper loses its wet strength within seconds of being saturated with water. With some wet-strength chemicals, the rate of wet-strength loss on soaking is slowed and paper made this way is then said to have temporary wet strength. Other additives give so-called permanent wet strength. However, such wet strength is never fully permanent, as both the resin and the paper have only a finite effective life [8].

7.2   Mechanisms of wet-strength development

To produce papers that retain some of their original dry strength when wet, it is necessary to do one of four things:

- add to or strengthen existing bonds
- protect existing bonds
- form bonds that are insensitive to water
- produce a network of material that physically entangles with the fibres

To achieve this, wet-strength additives have been developed which are, in general, chemically-reactive, water-soluble polymers. The chemical reactivity can be of two kinds; either the polymers can react with themselves (homo-crosslinking) or they can react with cellulose or the materials at the cellulose interfaces (co-crosslinking).

Wet-strength mechanisms have been reviewed by a number of authors [9–12] but, in general, two methods are invoked to explain the development of wet strength in paper [13]. The first is referred to variously as the 'preservation', 'restriction' or 'homo-crosslinking' mechanism [14] where wet-strength additives are adsorbed by cellulose and form a crosslinked network when the paper is dried. When the paper next comes into contact with water, rehydration and swelling of the cellulose is restricted by the resin network. Thus, a portion of the original dry strength is encased and preserved. The second mechanism has been referred to as a 'reinforcement', 'new bond' or 'co-crosslinking' mechanism. In this case, it is suggested that there is crosslinking of the fibres by the wet-strength resin. The bond then persists after any naturally occurring bonding has been destroyed by water. In this case, covalent crosslinking would lead to a stronger, more permanent form of wet strength, whereas ionic bonding would give a more temporary form. For clarity, the nomenclature of Westfelt [10] will be used, and the mechanisms will be referred to as homo-crosslinking and co-crosslinking.

In the following review of the chemistry of commercially available wet-strength additives, the mechanisms that are thought to operate will be described; however, in many cases, the mechanistic evidence in the literature is not clear.

7.3   The chemistry and application of commercial wet-strength resins

Since the series of discoveries, during the late 1930s and 1940s, that adding certain water-soluble synthetic resins to a fibre furnish and curing *in situ* brings about a sharp increase in wet strength, there has been a rapid growth in the use of wet-strength resins and wet-strength papers. In general, the resin must be highly dispersible in water (hydrophilic), reactive and of opposite charge to the fibre furnish i.e. cationic. These properties encourage adsorption of the resin. It is also imperative that the resin cures and develops wet strength, as there are many products, particularly those developed for the textile industry, that are unsuitable for paper use because of the high temperatures that are required for curing.

There is also a class of chemicals that are not water soluble but have been used to give wet strength to paper. These are the synthetic elastomers, often latexes, such as the polyvinyls, acrylics and styrene-butadienes. Important differences between these and the water soluble products are: firstly, they give rise to other properties in the paper when used in sufficient quantities to give wet strength; secondly, 5–40% more is required to give wet strength; and thirdly, the mechanisms by which they operate are different.

Of the numerous patents describing compositions for wet-strength chemicals, relatively few have been exploited and developed into products that have gained commercial acceptance. Initially, the aminoplast resins (urea–formaldehyde, melamine–formaldehyde) set the standard for cost, performance and convenience. Later, resins were developed which cured under neutral or alkaline conditions. The following sections review each of the major chemical groups of wet-strength additives taken essentially in chronological order of their appearance on the papermaking scene.

7.3.1 *Urea-formaldehyde resins*

The earliest process for producing wet strength was to heat paper to high temperatures or to parchmentise it in dilute sulphuric acid [15]. Later, experiments showed that the impregnation of paper with formaldehyde gave rise to wet strength, but the offensive odour and the brittleness of the paper resulting from its low pH were drawbacks [16]. Dimethylolurea, the reaction product of urea and formaldehyde, produces wet strength in paper, but

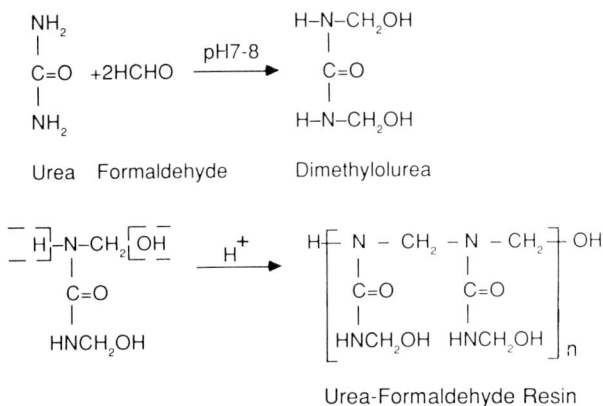

Figure 7.1 The preparation and structure of urea–formaldehyde resins.

because it is readily soluble in water and not substantive to pulp fibre, it is not effective as a wet-end additive. Therefore, attention was focussed on developing polymers of urea and formaldehyde and, initially, a water insoluble resin was produced by condensing dimethylolurea with itself (Figure 7.1). Although this material gave wet strength, its insolubility meant that it could only be used commercially as a tub-sizing additive [17].

The next advance was to modify the urea-formaldehyde (UF) resin with either sodium bisulphite or glycine to produce a water-soluble, anionic polymer which could be added at the wet end and precipitated onto pulp by the use of alum. These products were later supplemented by UF resins which were modified with diethylenetriamine or other polyfunctional amines to give a cationic resin which was substantive to pulp [18].

Figure 7.2 shows the structure of UF resin and indicates the homo-crosslinking reaction where the methylol on one polymer chain reacts with a nitrogen-bound hydrogen on another [19]. The result is an insoluble, three-dimensional network which can protect existing fibre–fibre bonds or restrict hydration and swelling when the paper is rewet.

UF resins are supplied as aqueous solutions that require no special pretreatment at the paper mill. The most widely used products contain 25–35% resin, although some solutions contain as much as 50% resin. On storing, these reactive resins continue to cross-link and increase in viscosity. As cross-linking progresses, the wet-strength performance improves until, ultimately,

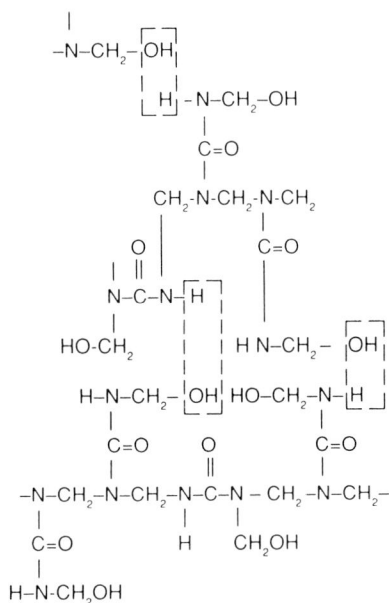

Figure 7.2 Homo-crosslinking of urea–formaldehyde resin.

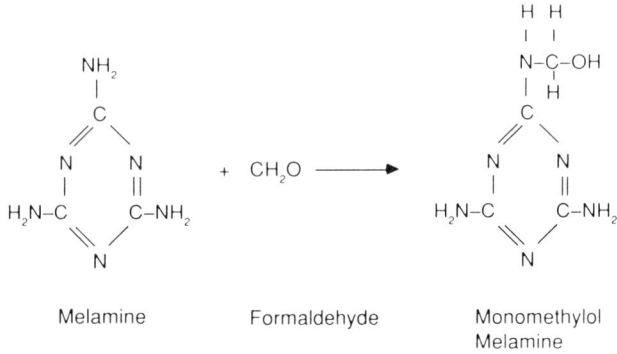

Figure 7.3 The first stage in the synthesis of melamine–formaldehyde resins.

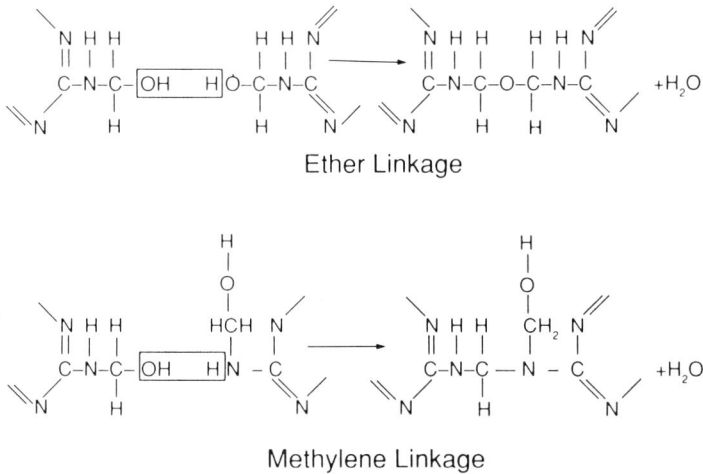

Ether Linkage

Methylene Linkage

Figure 7.4 The reactive groups of MF resins that lead to homo-crosslinking.

the reaction proceeds too far and a useless gel is formed. Therefore, in practice, a compromise is reached between efficiency and storage life. Dilution of the resin before use extends the storage life.

UF resins are thermosetting, and curing is accelerated by heat and particularly by a low pH. For example, the UF resin in paper made at a pH of 5.5 will eventually react to the same level of wet strength as paper made at pH 4.5 but it will require significantly longer.

7.3.2 *Melamine–formaldehyde resins*

The next development was the melamine–formaldehyde (MF) resins, which were first patented for use in paper in 1944 [20]. To prepare the resin, melamine is reacted with formaldehyde to give methylol melamine (Figure 7.3). Methylol derivatives ranging from monomethylol to hexamethylol melamine can be prepared by increasing the molar ratio of formaldehyde to melamine. Initially, however, the trimethylol derivative (TMM) was found to perform best [21]. Because monomeric derivatives tend to crystallise and become difficult to handle, most commercial wet-strength resins are made from the easier-to-handle syrups produced by condensing two or more monomer units. A number of reactions probably occur, but the formation of ether and methylene linkages are important in providing wet strength in paper. The crosslinking reactions which are promoted by high temperature and low pH [22] are shown in Figure 7.4.

Early work showed that aqueous solutions of TMM gave excellent wet strength when applied to finished paper, but cost and difficulty in curing made them uncompetitive with UF resins. The solution was to form water soluble salts by treating TMM with hydrochloric acid. The aqueous solutions of these salts polymerise by condensation to produce positively charged colloidal particles which are very substantive to negatively charged pulp fibres.

Although the trimethylol product was originally found to be the most effective, it was later shown that superior colloids could be prepared by adding an excess of formaldehyde to the TMM colloid in a two-stage process. Later it was found that a simple adjustment of the amount of acid gave the same product from a single ageing period and so the high efficiency (HE) products were developed. HE colloid preparations are more effective in any application, but are particularly tolerant of high multivalent ion concentrations such as the high sulphate levels found in hard water regions or where aluminium sulphate (papermaker's alum) is used. A range of HE colloids is produced commercially containing typically from 1 to 7 moles of formaldehyde in addition to the three in the basic TMM. Higher amounts of formaldehyde have occasionally been used. A detailed account of the chemistry of melamine acid colloids has been given by Maxwell [23].

In papers treated with MF resins, the wet strength only develops as the paper is dried with heat, and the available evidence suggests that the mechanism of wet-strength development is homo-crosslinking of the resin by the interaction of a methylol group on one polymer chain with a nitrogen-bound hydrogen on another [9]. This is analogous to the processes occurring with UF resins. The resulting three-dimensional polymer network is insoluble in water, thus promoting wet strength by a homo-crosslinking mechanism. There is some early evidence for co-crosslinking where it was shown that MF resin increased the wet and dry strength of paper by equal amounts [24]. Because of the relatively large number of reactive groups when compared to UF resin, paper produced with MF resin has greater wet strength per unit of resin used and greater permanence during ageing. On the other hand, UF resins are considerably cheaper. MF resins, like UF resins, require acidic papermaking conditions for best performance.

Although formaldehyde resins remained a commercially important class of products for many decades, regulatory compliance and health and safety concerns over formaldehyde emissions have significantly reduced the use of these products Major applications for MF resins are in providing permanent wet-strength in currency and map papers.

7.3.3 Epoxidised polyamide resins

The need for a wet-strength resin which could be used in neutral or alkaline papermaking led, over thirty years ago, to the development of poly-(aminoamide)-epichlorohydrin (PAE) resins [25, 26], sometimes referred to as polyamide–polyamine–epichlorohydrin (PPE) resins. The preparation of PAE resins is similar to that used in preparing 66 Nylon. A dibasic acid is condensed with diethylene triamine to give a water-soluble polyamide. The secondary amine groups of the polyamide are then alkylated with epichlorohydrin, and a number of reactions take place. Predominantly, tertiary amino-chlorohydrin groups are formed, which self-alkylate to form 3-hydroxyazetidinium groups which are responsible for the reactivity and the cationic character of this wet-strength resin (Figure 7.5). The azetidinium group will slowly react with water to form a non-reactive diol (Figure 7.6). Treatment with epichlorohydrin leads to some crosslinking and care has to be taken to control this and to retain water solubility. The resulting product consists of relatively low molecular weight polyamide backbones with many reactive side chains. Chloride counter-ions are formed as a result of the epoxidation and, at the end of the manufacturing process, the resin is usually acidified to provide stability against gelation during storage. The acidity and presence of chloride ions requires that high quality stainless steel or suitable plastic containers are used for storage. There have been a number of reviews of the chemistry of PAE resins [27, 28], the most recent and comprehensive being that of Espy [29] which also covers the polyalkylene polyamine–epichlorohydrin (PAPAE)

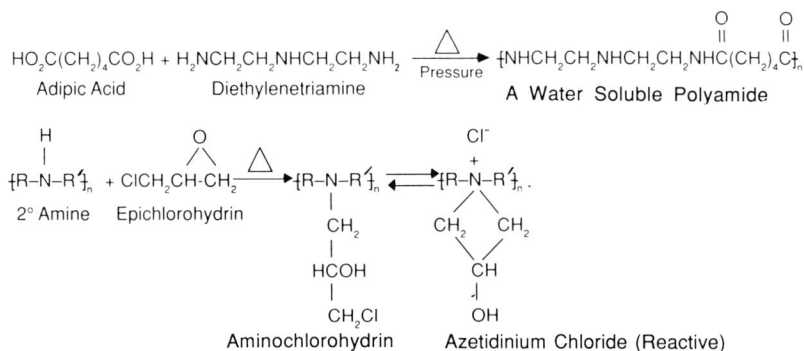

Figure 7.5 The preparation of poly(aminoamide)-epichlorohydrin resins.

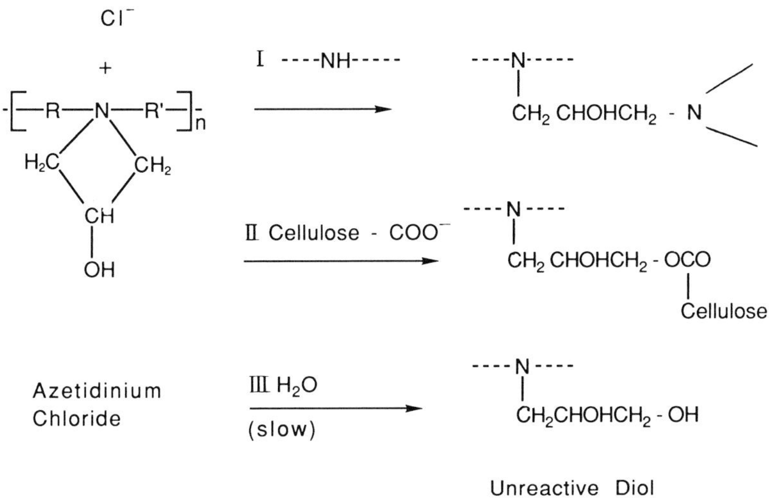

Cl⁻

```
        +                I  ----NH----        ----N-----
                                                   |                    /
  -[-R—N—R'-]ₙ    ──────────────▶          CH₂ CHOHCH₂ - N
      / \                                                            \
    H₂C   CH₂
      \   /
       CH
       |
       OH
                                               ----N----
                          II Cellulose - COO⁻        |
    ──────────────▶          CH₂ CHOHCH₂ - OCO
                                                     |
                                                 Cellulose


  Azetidinium        III H₂O                  ----N----
  Chloride           ─────────▶                    |
                     (slow)              CH₂CHOHCH₂ - OH


                                            Unreactive  Diol
```

Figure 7.6 The reactivity of poly(aminoamide)-epichlorohydrin resins. I: reaction through the amine groups of other resin molecules (homo-crosslinking); II: reaction with carboxyl groups in cellulose (co-crosslinking); III: slow reaction with water.

and 'amin–polymer'–epichlorohydrin (APE) resins that were later introduced to the paper industry but have not commanded the same volume of use in the wet-strength chemical market.

As the resin is dried in a paper sheet, the hydroxyazetidinium groups have two possibilities for reaction. One is to react through the amine groups of other resin molecules to give a homo-crosslinked network, and the other is to react with available carboxyl groups in the paper (Figure 7.6). There is some evidence suggesting that PAE resins are predominantly involved in homo-crosslinking rather than being cellulose-reactive.

Earlier literature [30] has outlined the development of an understanding of PAE wet-strength mechanisms. It has been shown that, when different pulp furnishes are treated with PAE resin and used to make unsized sheets covering a wide range of dry strengths, similar percentages of wet to dry strength are obtained [13]. This suggests that the pulp is not directly involved in the reactions responsible for wet-strength development. It has also been shown that, in handsheets made at high PAE resin addition levels, the incremental wet and dry strengths depended on pulp beating in a way that is characteristic of other homo-crosslinking resins [13]. Earlier, it was shown that, when using methyl-β-glucoside as a model for the anhydroglucose units of cellulose, PAE resin did not react appreciably with hydroxyl groups under typical paper machine drying conditions [31]. However, this model did not take into account that cellulose and hemicellulose have an appreciable amount of carboxyl groups in addition to hydroxyls. Following on from this work, Espy and Rave [30] developed a procedure using cupriethylenediamine

(CED) as a tool for detecting the permanent crosslinking of cellulose by alkaline curing resins. They concur with Bates [31] that PAE resin does not readily react with hydroxyl groups to form ether linkages. But their work goes on to suggest that PAE resins first wet-strengthen paper by forming CED-labile ester links with pulp carboxyl groups and then form additional resin homo-crosslinks instead of reacting with available cellulosic hydroxyls.

The wet strength is not fully developed by the time a paper product reaches the end of a paper machine, and strength continues to develop as the paper is stored. The rate of curing is affected by the temperature and the length of time in contact with heat [28]. The bonds that are eventually formed are not readily broken under either acid or slightly alkaline conditions, and the wet strength is therefore permanent. Whilst this can be an advantage in certain applications, it does make the paper or board more difficult to repulp. Details of the approaches used to repulp are given by Pahl and Espy [32]. Essentially, it is necessary to use sodium hypochlorite at pH 6.5–7 (hypochlorous acid being the dominant species); a temperature of 50–55°C; and a mechanical disintegrator for bleached pulps. For unbleached systems, a high pH (11–12 using sodium hydroxide), and a high temperature (77°C), in addition to mechanical treatment, are usually necessary.

The practical aspects of using PAE resins in papermaking have been reviewed by Pahl and Espy [32], and only some of the more important points will be highlighted. Being highly charged cationic polymers, PAE resins are substantive to negatively charged fibres and are rapidly adsorbed when added to a papermaking furnish. At low dosages there is a steady increase in wet strength, but a point is reached where resin retention starts to decrease as the adsorption sites on the fibres (carboxyl groups) diminish and furnish becomes less negatively charged. It is possible to completely reverse the charge of some furnishes, although the dosages required to do this are usually uneconomic. The carboxyl content of the pulp affects performance of the resin, and typically a higher wet strength can be achieved with unbleached kraft than bleached kraft. Bleached sulphite pulps (which have a lower carboxyl content) are the most difficult to treat. As would be expected, refining a pulp produces an increase in surface area and this affects performance. At low resin dosages the effect is not evident, as even an unrefined pulp provides sufficient surface for resin adsorption. Where the effect is seen, is at higher dosages when more resin can be adsorbed by the refined pulp. A detailed study of the effects of refining on wet-strength response of the most important classes of wet-strength additives has been reported by Espy [13].

There is a synergy between PAE resins and the water soluble anionic polymer carboxymethylcellulose (CMC), and in many practical applications the two are used together. More wet and dry strength can be developed than when the resin is used alone, and additional dry strength is obtained which can be advantageous to the papermaker (see section 7.2.4). The theory is that retention of PAE resin is limited because, with increasing additions of resin, the anionic sites on the pulp (carboxyl groups) become saturated and ultimately a point is reached where the fibres are cationic and therefore of the same charge as the resin. Adding CMC improves resin retention and allows more resin to be added before this plateau in performance is reached. The mechanism by which the CMC functions is complex but can be thought of as follows: there is an optimum in the CMC/resin ratio as the CMC initially interacts with the resin to form a neutral to slightly cationic complex. However, if the optimum of this ratio is exceeded, then the formed complex is anionic and poorly adsorbed. In general, the optimum CMC/resin ratio is in the region of 0.4 to 1.0.

PAE resins are most often used in a pH range of 6 to 8, but their efficiency is adequate in the pH range 5 to 9. Reducing the pH of a papermaking furnish leads to two effects that discourage adsorption. Firstly, fewer carboxyl groups on the pulp are dissociated. Secondly, although most of the charge on the PAE resin is due to the azetidinium groups, which are insensitive to pH change, there are some tertiary amine groups present in the polymer and these lose their charge at lower pH.

PAE resins are sensitive to anionic interfering substances whether they are naturally occurring (e.g. lignosulphonates) or anionic chemical additives (e.g. anionic dyes used to dye tissue). In both cases, it is necessary either to add more resin to obtain the required effect, or to use a high charge density cationic polymer to pretreat the pulp [33].

The effect of dissolved inorganic ions in the papermachine system is less clear. Low levels (< 100 ppm calcium) can improve performance, but higher levels lead to a decrease in wet strength [32]. Presumably, cationic ions compete with the resin for the available carboxyl groups. In a detailed study of the effects of inorganic ions, Ampulski and Neal [34] have shown that PAE adsorption decreases as the valence of the metal ion in solution increases, and that the quantity of adsorbed PAE resin which then binds to the pulp decreases as the salt concentration in solution increases. PAE resins are also very sensitive to chlorine chemicals and, if it is not possible to exclude these chemicals after pulp-brightening or microbial control, it is usually necessary to add an antichlor agent such as sodium sulphite before adding the resin.

Today, PAE resins are used extensively in practically all types of wet-strength papers. Major examples include: household products such as paper towels, napkins, and facial tissue; packaging materials, for example liquid packaging, corrugated boxes, and paper bags; and specialties such as photographic papers, wrappings and disposables. PAE resins are usually recommended for coffee filter papers. Because of their quenching effect they are not recommended for use with fluorescent whitening agents.

7.3.4 *Glyoxalated polyacrylamide resins*

In the late 1960s, a glyoxalated polyacrylamide resin was introduced to the paper and board industry [35] and it has now become widely accepted. Unlike the epoxidised polyamide resins, this resin gives temporary wet strength to paper and behaves as a dry-strength additive. Such a combination can be desirable in certain paper products. Although it has a variety of applications, its principal use is in tissue grades and, particularly, in wet-strength towelling.

Glyoxalated polyacrylamide is prepared by crosslinking a low molecular weight polyacrylamide (PAM) with glyoxal. The PAM is normally a copolymer of acrylamide and a quaternary ammonium cationic monomer which is prepared in aqueous solution (Figure 7.7). This results in a cationic polymer which is substantive to pulp fibre, although the PAM can be anionic, in which case a retention aid would have to be used in the application. The cationic backbone is then crosslinked with sufficient glyoxal to react with most, but not all, of the PAM backbone amide groups. On storage, the resin continues to crosslink and can ultimately gel, therefore in order to achieve the desired stability, paper mills typically dilute the resin on receipt. At 25°C, a 10% solution will gel in about 8 days, whereas a 6% solution stored under the same conditions will take about 65 days to gel [36].

There is strong evidence to suggest that glyoxalated PAM imparts wet strength primarily through covalent bond formation between resin and paper fibres (co-crosslinking). This bonding is by hemiacetal formation involving the addition of cellulose hydroxyl to free aldehyde groups on the resin [37] as shown in Figure 7.8. That there is some intermolecular crosslinking within the resin can be taken for granted but, in order to function, there need to be at least some fibre–resin–fibre bonds within the fibre–fibre bonded area. It is also thought that a certain amount of bond protection occurs where water is prevented from immediate access to the water sensitive interfibre bonds.

The reaction of glyoxalated PAM with cellulose is rapid at neutral pH (6–8) and even more rapid at an acidic pH (4–6), giving 80–100% of the potential

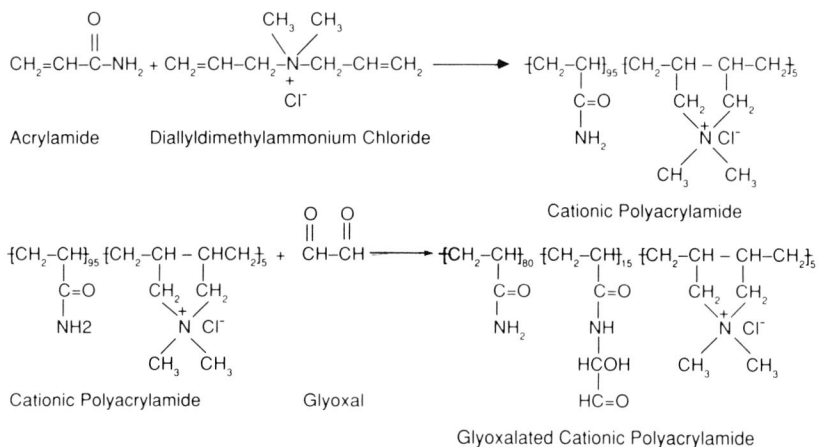

Figure 7.7   The preparation of glyoxalated polyacrylamide.

Homo-Crosslinking

$\{CH_2-CH\}_n + \{CH_2-CH\}_n$ $\xrightarrow{H^+}$ $\{CH_2-CH\}_n$

C=O	C=O	C=O
NH$_2$	NH	NH
	HCOH	HCOH
	HC=O	HOCH
		NH
		C=O
		[CH2–CH]$_n$

Acrylamide Gloxalated Acrylamide   Amidol

Co-Crosslinking

$\{CH_2-CH\}_n + 2\text{Cell-OH}$ $\underset{H_2O}{\rightleftharpoons}$ $\{CH_2-CH\}_n$ $\xrightarrow{H^+}$ $\{CH_2-CH\}_n$

C=O	C=O	C=O
NH	NH	NH
HCOH	H–C–OH	H–C–OH
HC=O	H–C–O–Cell	H–C–O–Cell
	OH	O

Gloxalated Acrylamide   Cellulose Hydroxyl   Hemiacetal Bond   Cell Acetal Bond

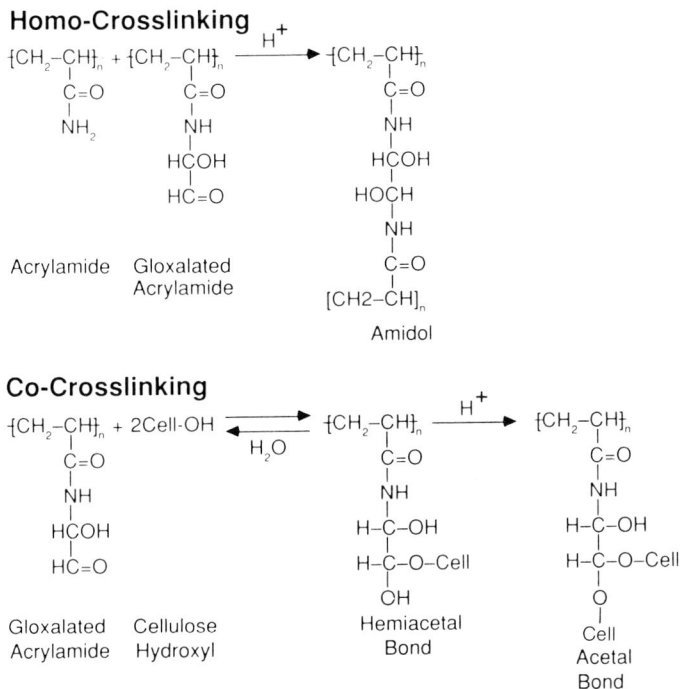

Figure 7.8 Crosslinking reactions of glyoxalated polyacrylamide [12].

wet-strength development on the paper machine. Compared with this, ageing or curing of the paper gives little or no additional wet strength. The reaction is reversible in the presence of water and a resin-treated paper gradually loses a portion of its wet strength on prolonged soaking. It has been shown that if a laboratory handsheet is soaked for four hours it still retains over 85% of its wet tensile strength after drying [36]. The loss in wet strength is gradual and therefore the wet strength is present during the time that temporary wet-strength products are normally used (e.g. paper towels for wiping up spills or drying one's hands). This temporary wet strength is also of sufficient duration to provide strength to a paper sheet passing through a size press or coater.

As would be expected from the chemistry of this bonding mechanism, it is possible to use conditions that affect reversibility and destroy the wet strength that is introduced. Two relevant examples can be considered. First, wet strength is destroyed by strong alkalis such as sodium hydroxide or hypochlorite solutions. The action of the alkali is virtually immediate and complete, and is irreversible, possibly due to the destruction of the aldehyde functionality. This has certainly not gone unnoticed by those using some of the alkaline household cleaning products with temporary wet-strength towelling. Second, sulphite or bisulphite ions can react to form the bisulphite adduct.

Although the addition of bisulphite has been suggested as a way of increasing the storage stability of glyoxalated PAM resins by inhibiting crosslinking [37], there has been no indication that this has been practised. This becomes important when sulphites are present in the paper machine wet end, from either residual brightening chemicals (hydrosulphite) or addition of antichlor. Concentrations of over about $2 \, mg/dm^3$ reduce resin efficiency. The bisulphite adduct is anionic and can offset part, or all, of the cationic charge of the resin, resulting in poor retention. In such cases, the addition of a small amount of a high charge-density, cationic polymer usually restores efficiency.

One of the properties of glyoxalated PAM resins is that they also contribute significantly to the dry strength of treated paper. Dry strength, which is due to the unreacted amide groups on the resin, is largely independent of wet-strength chemistry, and is typical of that obtained with conventional polyacrylamide dry-strength resins [39, 40]. The dry strength contributes to sheet stiffness, which is desirable in certain napkin and speciality crêped paper grades, but can be undersirable where softness and feel are important (e.g. tissue). Usually, the added dry strength is taken advantage of indirectly, as a means of reducing cost or improving quality, where dry strength is the limiting factor. Examples of these advantages are increased bulk, increased use of weaker fibre and basis weight reduction.

A more detailed practical guide to the use of glyoxalated PAM resins can be obtained by referring to Farley [36, 41].

7.3.5 *Comparison of properties of the predominant commercial resins*

These chemistries form the bulk of the commercial market for wet-strength resins supplied to the paper industry. In Table 7.1 the major properties and attributes of the resins are compared and summarised. There is a risk in preparing a table such as this, as there are always likely to be exceptions, and therefore this is intended only to be a useful but general guide.

7.3.6 *Polyethylenimine resins*

Polyethylenimine (PEI) was first introduced to the paper industry in the late 1940s in Germany (IG Farbenindustrie AG, German Patent No. 758570). The original process for making the ethylenimine monomer consisted of converting ethanolamine to aminoethyl hydrogen sulphate by treatment with sulphuric acid. The aminoethyl hydrogen sulphate was then treated with sodium hydroxide to give ethylenimine. Ethylenimine, both in bulk and in solution, readily polymerises to PEI by the action of acid catalysts [42] as shown in Figure 7.9. With appropriate control of the reaction conditions, it is possible to produce products over a wide range of molecular weights (600–50 000 g/mol) and it is generally agreed that the molecules are predominantly linear but contain branches whose frequency depends on the mode of polymerisation.

Table 7.1 Comparison of wet-strength resins

Resin	MF	UF	Glyoxal/PAM	PAE

Principal advantages	Permanence
	Low cost
		Low cost
		Easier
		  repulping
			Neutral sheet
			Dry strength
			Repulping
			Temporary
				Neutral or
				  alkaline
				Retention
				Permanence

On-machine cure (%)	50–60	20	60–95	10–30
Time to 100% cure (weeks)	1–2	1–4	1–2	1–2
pH range	4.0–5.5	3.8–4.5	4.5–7.5	5.0–9.0
Best operating pH	4.5	4.0	6.0–7.0	8.0
Solids content (%)	12% (regular) 10% (HE)	25–40	6–10	12–33

Point of addition	Usually
	  thick stock
		Thick stock	Thick (or
			  dilute)
				Thick
				  (or dilute)

First pass retention (%)	60	35	40	80
Major precaution	Sulphate level	Low pH	Sulphites pH over 7.5	Chlorine chemicals

Broke handling	High temperature
	Low pH
		Easier than MF	Easiest
			Helped by high
			  temperature
			  and high pH
				Hypochlorite
				  or high
				  pH and
				  temperature

Absorbency (without rewetters)	Poor	Fair	Best	Good
Sheet brightness	Fair	Good	Best	Fair
Usual drainage	Slower	No effect	Slightly faster	Faster

Storage (24°C) (weeks)	1	12–24
		  at lower
		  solids
			1 (10%)
			4 (7.5%)
				12

Relative cost	53	29	100	100

Affect on sizing	Much
	  improved
		Little	Improved	Much
				  improved

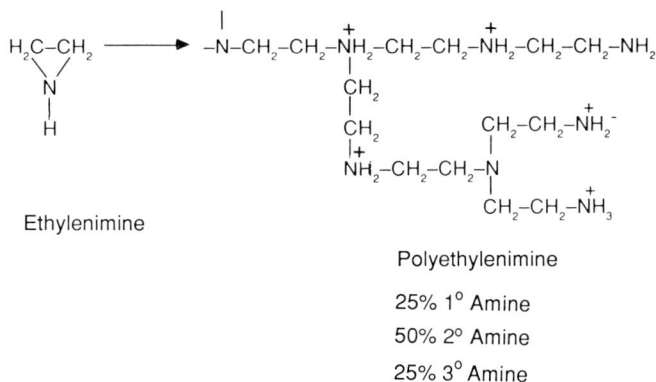

Figure 7.9 Preparation and chemical structure of polyethyleneimine (PEI).

The mechanism by which PEI works is somewhat different from the resins that have been discussed in earlier sections: (i) wet strength is developed without having to cure the paper; (ii) the level of wet strength that can be attained is less than with the thermosetting resins; and (iii) continued soaking, especially under acidic conditions, causes loss of wet strength. It has been proposed, therefore [11] that PEI functions by creating stronger, interatomic bonding, rather than by forming homo- or co-crosslinked networks [43, 44]. Work by Allan and Reif [45] has certainly helped in confirming the view of Trout [44] that PEI forms an ionic bridge between cellulose carboxyl groups and the cationic groups in the resin that is sufficiently strong to overcome any interaction that might occur between water and the polymer. The evidence comes from experiments that were done to examine the influence of the anionic charge density of fibres on the wet strength of PEI treated paper. The number of anionic sites was increased by adsorbing a dye containing anionic sulphonate groups onto the cellulose. Wet strength was enhanced, suggesting that PEI appears to co-crosslink, with the bonding being ionic in nature. The amine cationic groups responsible for wet strength have dissociation constants (pK) of around 12 and it has been shown that the retention and performance of PEI is best in the pH range of 7–9.

There are a number of reasons why PEI has not been more extensively used as a wet-strength resin in the paper industry. First, in the past, it has proved to be more expensive than the thermosetting resins and this is certainly related to the cost of manufacture of PEI and the availability of the ethylenimine monomer. Second, the use of PEI in quantities sufficient to give wet strength causes yellowing and loss of brightness and therefore it is restricted to use in off-white or dyed products. Quaternising PEI does remove the discolouration effect which suggests something about the mechanism of the colour formation, but its ability to give wet strength is also removed by this treatment [42]. A more thorough review of the chemistry and use of PEI resins has been given by Sarkanen [42].

7.3.7 *Dialdehyde starch*

Another wet-strength additive that has found use in the paper industry is dialdehyde starch (DAS). It is essentially a highly modified starch in which the vicinal hydroxyl groups (at the C-2 and C-3 carbons) are selectively attacked by periodic acid severing the C-2 to C-3 bond to form dialdehyde starch (Figure 7.10). This selective oxidation can be controlled and in most cases 80–90% conversion to dialdehyde starch is found. The aldehyde groups are not present to any extent as free aldehydes, but rather as hemiacetals or as hemialdals [46]. Since the linkages in these compounds are weak, dialdehyde starch reacts as if the aldehydes were free, permitting its use as a reactive polyaldehyde capable of reaction through hydroxyl, amino, or imino groups. This property makes it of potential value in crosslinking cellulosic hydroxyls

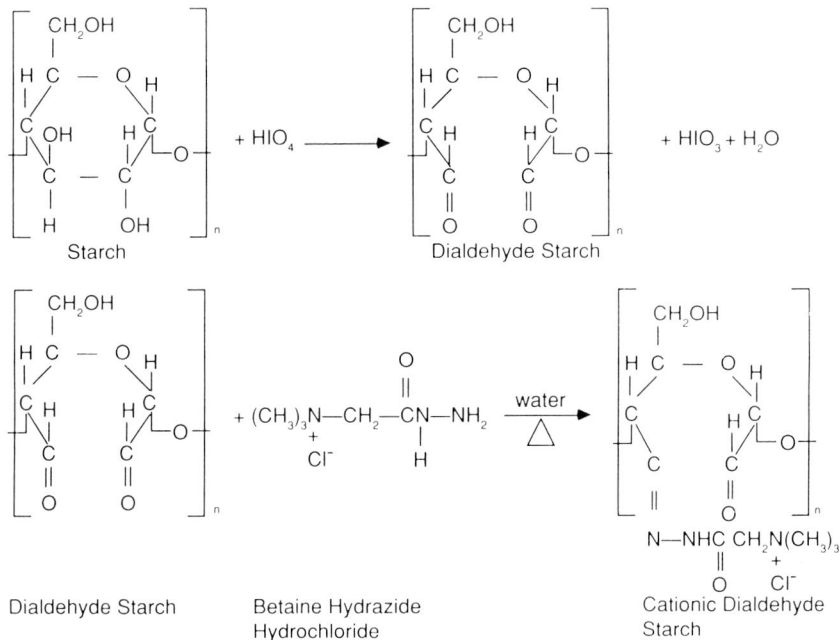

Figure 7.10   The preparation of dialdehyde starch and its modification into a cationic product.

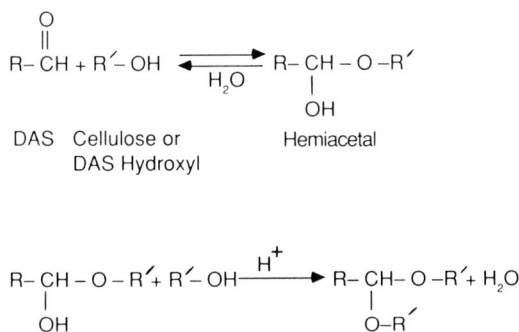

Figure 7.11   Crosslinking reactions of dialdehyde starch.

in paper to give temporary wet strength [47]. DAS is supplied in the form of a fine powder, similar to the original starch. It can then be further modified in a number of possible ways to give a cationic or anionic product. Cationic DAS has been found to be particularly effective in this regard because of its affinity for cellulosic pulp [48]. For example, reaction with betaine hydrazide hydrochloride (Figure 7.10) in the presence of water and heat gives a cationic DAS betaine hydrazone [12, 49].

Wet strength from DAS is due to co-crosslinking of DAS aldehyde groups and cellulose hydroxyls to form hemiacetal or acetal bonds [50]. High pH favours the formation of hemiacetal bonds which are temporary and easily broken by water. Low pH encourages the DAS/cellulose reaction to proceed to the formation of acetal bonds which are more permanent. Their hydrolysis requires more acidic conditions, which are not found in tap water. DAS can be applied in the pH range of 4.5–6.5. The crosslinking reactions of DAS are shown in Figure 7.11. As would be expected from starch-based products, DAS gives high dry strength.

7.3.8   Other wet-strength resins

Apart from the specialised application of synthetic latexes, no other wet-strength resins have gained wide commercial use. For more detail, the chapter by Britt [11] or the monograph edited by Weidner [51] should be consulted. A partial list of other wet-strength resins is as follows:

- proteinaceous adhesives treated with formaldehyde
- cellulose xanthate (viscose)
- synthetic latexes
- vegetable gums
- glyoxal

Not all are substantive to fibre, therefore they may have to be applied to the surface of paper or augmented in some way to encourage retention.

Recently it has been reported [52] that block reactive group starches are now commercially available to the paper industry. A starch is modified so that it is cellulose-reactive. Through chemical reaction, the reactive groups are then blocked to stabilise the product until it is ready to be used. In use, the starch is cooked and the blocking groups removed leaving the starch to react with the cellulose fibres. This product is a temporary wet-strength additive which also contributes, as would be expected, to dry strength.

7.4   Testing of wet-strength papers

The obvious way to determine wet strength is to measure its burst or tensile strength when wet, but there are a number of important variables that have to be considered if a meaningful result is to be obtained. Essentially, the test should relate to the end use of the wet-strengthened paper. The requirements of a test to show the performance of a resin in a coffee filter paper are obviously very different from those used to test a consumer towelling product. Variables of importance include the type of liquid that the paper is likely to be in contact with, its pH, temperature and ionic strength, and the contact time. Nevertheless, there are useful Standard Methods for the determination of wet strength (e.g. TAPPI Method T456-os-68), although many non-standard tests have been developed over the years.

In the Standard TAPPI Method, a strip of paper is completely wetted before applying a breaking force. The paper is immersed in water or, if it is too weak, it is mounted in the jaws of a tensile tester and wet midway over a distance of 2.54 cm. The load required to break the paper is then recorded. Wet strength values alone are of little use; therefore the dry strength is always measured, and the result reported as the wet:dry ratio or percent wet strength (wet strength as a percentage of the dry strength). Other common paper tests such as burst, tear, rub resistance and wax pick are also used, depending on what effect the papermaker is looking for from the wet-strength resin. More information on the testing of wet-strength papers can be obtained from Britt [11] or Allison [53].

7.5 Assessing the efficiency of wet-strength resins

Since a papermaker must provide a certain test value in his product, the efficiency of a wet-strength resin is most meaningfully defined as the amount of additive required to give a desired result. This is because the dose-performance relationships of resins of different chemistries or resins used under different conditions are not linear. Using the rationale of Farley [36], Figure 7.12 shows the dose-performance curves drawn so that the wet strength

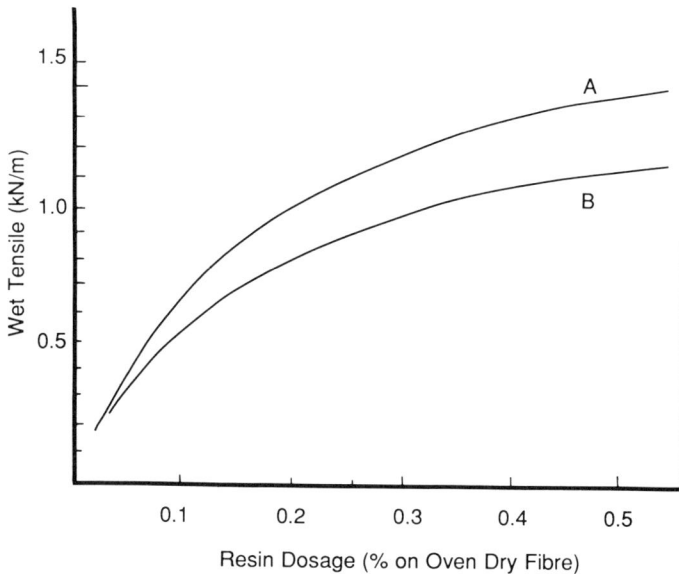

Figure 7.12   The dose response curve for two wet-strength additives.

of curve B is 80% of that of curve A at all doses. To achieve a wet-tensile strength of 1.0 kN/m required 0.28% A or 0.55% B. Thus the efficiency of B relative to A at this test level is only around 51%. But the papermaker may complain that the use of B requires a 96% higher treatment level than A. These large increases in resin requirement should be kept in mind because what may seem to be a rather modest loss in performance can translate into a large increase in resin usage and costs. The converse, of course, is also true.

7.6   Future trends

In this age of environmental awareness, some of the current classes of wet-strength additives have come under scrutiny. Particular examples are the for-maldehyde-containing aminoplast, and the PAE resins. The latter have been shown to contribute to adsorbable organic halogen (AOX) emissions from paper mills. This realisation has arisen mainly from the developments in sophisticated analytical instrumentation which now allows the extremely sensitive detection and analysis of chemicals. The chemical industry has reacted responsibly, and committed a considerable and successful research and development effort. UF resins have been produced with lower 'free' formaldehyde [54, 55], and the industry has increased efforts to reduce workers' exposure to formaldehyde [56]. Research on PAE resins has focussed on reducing AOX [57] and, in particular, on reducing the concentrations of the by-products such as 1,3-dichloro-2-propanol and 3-chloro-1,2-propanediol. This effort will continue.

There will continue to be research into new wet-strength chemistries. If the chemical companies were in a position to reveal proprietary information it would be seen that a tremendous number of molecules have been synthesised and tested for wet-strength activity in paper. Many of these have been shown to perform, but have not been commercialised because of other factors. Many sit on shelves waiting for a market niche or an opportunity when they would be cost effective. A recent example is the growing number of publications on the potential use of vinyl amine copolymers as wet-strength resins [58] — this approach is limited by the current high cost of producing vinyl amine.

Environmental pressures will continue to influence research into new wet-strength chemicals and any new legislation should create opportunities for chemical suppliers. There is a need for new products that are more biodegradable and more compatible with the ecologies of all areas of the environment with which they could possibly come into contact. The cationic resins are, by design, rapidly adsorbed by the materials in a paper furnish and are usually removed with the paper. This is desirable, but care has to be taken to avoid any high concentrations where the only surfaces available for adsorption are those of the natural flora and fauna (e.g. the gills of fish). High concentrations are avoided by shipping diluted products but this adds to the cost; therefore, there is a need to develop biodegradable, environmentally-compatible products which could be transported at high concentrations. A more detailed discussion of the trends and needs for the future has been given by Pelzer *et al.* [55, 59].

References

1. Britt, K.W., 'Review of developments in wet-strength paper'. *Tech. Assoc. Papers*, **31** (1948), 594–596.
2. Salmen, N.L., 'Mechanical properties of wood fibers and papers', in *Cellulose Chemistry and its Applications*, eds Nevell, T.P. and Zeronian, S.H., Ellis Horwood, Chichester (1987), Ch. 20, 505–530.
3. Back, E.L., 'Cellulose at high temperatures. Autocrosslinking, glass transition and melting caused by laser rays'. *Papier*, **27** (10A) (1973), 475–483.
4. Robertson, A.A., *Pulp and Paper Report No 26*, Pulp and Paper Research Institute of Canada, Montreal, Canada (1969).
5. Britt, K.W., *Tappi J.*, **46** (12) (1963), 154A.
6. Thode, E.F. and Guide, R.G., 'A thermodynamic interpretation of the swelling of cellulose in organic liquids: the relations among solubility parameter, swelling and internal surface'. *Tappi J.*, **42** (1959), 35–39.
7. Britt, K.W., 'Some observations on wet-strength paper'. *Paper Ind. Paper World*, **26** (1) (1944), 37–41, 46.
8. Maxwell, C.S. and Reynolds, W.F., 'Permanence of wet-strength paper', *Tappi J.*, **33** (4) (1950), 179–182.
9. Stannett, V.T., 'Mechanisms of wet-strength development in paper', in *Surfaces and Coatings Related to Paper and Wood*, eds Marchessault, R.H. and Skaar, C., Syracuse University Press, Syracuse, New York (1967), Ch. 10, 269–299.
10. Westfelt, L., 'Chemistry of paper wet strength. I. A survey of mechanisms of wet-strength development'. *Cellulose Chem. Technol.*, **13** (1979), 813–825.
11. Britt, K.W., 'Wet-strength', in *Pulp and Paper Chemistry and Chemical Technology*, 3rd Edition, ed. J.P. Casey, John Wiley & Sons, New York (1980), vol. III, Ch. 18, 1609–1626.
12. Neal, C.W., 'A review of the chemistry of wet-strength development', in *1988 Tappi Wet and Dry Strength Short Course*, Chicago, April 13–15 (1988), Tappi Press, Atlanta, 1–24.
13. Espy, H.H., 'The effects of pulp refining on wet-strength resin'. *Tappi J.*, **70** (7) (1987), 129–133.
14. Fredholm, B., Samuelsson, B., Westfelt, A. and Westfelt, L., *Cellulose Chem. Technol.*, **17** (3) (1983), 279.
15. Schur, M.O., *Improving the Wet-Strength of Papers*. US Patent 2,116,544, (1938) Brown Company, 10.05.1938.
16. Schaeffer, A., 'Chemical processes in the treatment of staple nylon with formaldehyde', *Zelwolle, Kunsteide, Seide*, **48** (1943), 1–8.
17. Britt, K.W., 'Review of developments in wet-strength paper'. *Paper Mill News*, **70** (39) (1947), 106, 109.
18. Yost, R.S. and Auten, R.W., US Patent 2,407,599 (1956), 17.04.56.
19. Brent, E.A., Drennen, T.J. and Shelley, J.P., 'Urea–formaldehyde resins', in *Wet Strength in Paper and Paperboard*, ed. Weidner, J.P. Tappi Monograph Series No. 29, Tappi Press, New York (1965), Ch. 1, pp. 9–19.
20. Woehnsiedler, H.P. and Thomas, W.M., *Cationic Melamine–Formaldehyde Resin Solutions.* US Patent 2,345,543 (1944), American Cyanamid Company, 28.03.1944.
21. Maxwell, C.S., *Pacific Pulp Paper Ind.*, **17** (4) (1943), 6.
22. Woehnsiedler, H.P., 'Urea–formaldehyde and melamine–formaldehyde condensations', *Ind. Eng. Chem.*, **44** (11) (1952), 2679–2686.
23. Maxwell, C.S., 'Melamine formaldehyde', in *Wet-strength in Paper and Paperboard*, Tappi Monograph Series No. 29, Tappi Press (1965), Ch. 2, 20–32.

24. Salley, D.J. and Blockman, A.F., *Paper Trade J.* **125** (1) (1947), 35.
25. Keim, G.I., US Patent 2,926,116, (1965), *High Wet-strength Paper*. US Patent 2,926,116, Hercules Powder Company, 23.02.1960.
26. Keim, G.I., US Patent 2,926,154, *Cationic Thermosetting Polyamide–Epichlorohydrin Resins for Preparing Wet-strength Paper*, Hercules Powder Company, 23.02.1960.
27. Moyer, W.W. and Stagg, R.A., 'Polyamide–polyamine–epichlorohydrin resins, in *Wet-strength in Paper and Paperboard*, Tappi Monograph Series No. 29, Tappi Press (1965), Ch. 3, 33–37.
28. Chan, L.L., 'Expoxidized polyamide resins', in *1988 Tappi Wet and Dry Strength Short Course*, Chicago, April 13–15 (1988), Tappi Press, Atlanta, 25–30.
29. Espy, H.H., 'Alkaline-curing polymeric amine-epichlorohydrin resins', in *Wet-Strength Resins and Their Application*, ed. Lock L. Chan, Tappi Press, Atlanta, Georgia (1994), Ch. 2, 13–44.
30. Espy, H.H. and Rave, T.W., 'The mechanism of wet-strength development by alkaline-curing amino polymer-epichlorohydrin resins'. *Tappi J.*, **71** (5) (1988), 133–137.
31. Bates, N.A., 'Polyamide–epichlorohydrin wet-strength resin II. A study of the mechanism of wet-strength development in paper'. *Tappi J.*, **52** (6) (1969), 1162–1168.
32. Pahl, B.L. and Espy, H.H., 'Use of polyamide resins in board', in *1988 Tappi Wet and Dry Strength Short Course*, Chicago, April 13–15 (1988), Tappi Press, Atlanta, 31–37.
33. Gill, R.I.S., 'CARTAFIX – An effective control of interfering substances', in *Specialty Chemicals for the Paper Industry, Proc. Pira Paper and Board Division Conference*, ref: 31/003/CM/81, Leeds, 6th March (1990), Paper No. 03.
34. Ampulski, R.S. and Neal, C.W., 'The effect of inorganic ions on the adsorption and ion exchange of kymene 557H by bleached northern softwood kraft pulp.' *Nordic Pulp Paper Res. J.*, **2** (1989), 155–163.
35. Coscia, A.T. and Williams, L.L., US Patent 3,556,932. *Water Soluble, Glyoxalated, Vinylamide Wet-Strength Resin and Paper Made Therewith*, American Cyanamid Company (1968).
36. Farley, C.E., 'Glyoxalated polyacrylamide wet-strength resin', in *Wet Strength in Paper and Paperboard*, Tappi Monograph Series, Tappi Press, Atlanta, (1991), in press.
37. Farley, C.E., Glyoxalated polyacrylamide wet-strength resin, in *1988 Tappi Wet and Dry Strength Short Course*, Chicago, April 13–15 (1988), Tappi Press, Atlanta, 39–42.
38. Coscia, A.T. and Williams, L.L., US Patent 3,556,933, American Cyanamid Company (1971).
39. Farley, C.E., in *Proc. 1986 Tappi Papermaker's Conference*, Tappi Press, Atlanta (1986), 147.
40. Wasser, R.B., in *Pulp and Paper Chemistry and Chemical Technology*, 3rd edn, ed. J.P. Casey, John Wiley & Sons, New York (1980), vol. III.
41. Farley, C.E., 'Glyoxalated polyacrylamide resin', in *Wet-Strength Resins and Their Application*, ed. Lock L. Chan, Tappi Press, Atlanta, Georgia (1994), Ch. 3, 45–61.
42. Sarkanen, K.V., 'Polyethylenimine resins', in *Wet Strength in Paper and Paperboard*, Tappi Monograph Series No. 29, Tappi Press (1965), Ch. 4, 38–49.
43. Wilfinger, H., *Papier*, **2** (1948), 265.
44. Trout, P.E., *The Mechanism of the Improvement of the Wet Strength of Paper by Polyethylenimine*, Ph.D. Thesis, Lawrence College (1951).
45. Allan, G.G. and Reif, W.M., 'Fibre surface modification Part 6. The Jack-in-the-box effect: a new mechanism for the retention of polyethylenimine and other polyelectrolytes by pulp fibres.' *Svensk Papperstidn.*, **74** (2) (1971), 25–31.
46. Mehltretter, C.L., 'Some landmarks in the chemical technology of carbohydrate oxidation'. *Die Stärke*, **15** (1963), 313.
47. Mehltretter, C.L., 'Recent progress in dialdehyde starch technology,' *Die Stärke*. **18** (1966), 208.
48. Hofreiter, B.T., Heath, H.D., Ernst, A.J. and Russell, C.R., Dialdehyde starch, an alkali labile wet-strength agent, *Tappi J.*, **57** (8) (1974), 81.
49. Hamerstrand, G.E., Hofreiter, B.T., Kay, D.J. and Rist, C.E., Dialdehyde starch hydrazones–cationic agents for wet-strength paper. *Tappi J.*, **46** (7) (1963).
50. Hofreiter, B.T., 'Dialdehyde starches', in *Wet-strength in Paper and Paperboard*, Tappi Monograph Series No. 29, Tappi Press, (1965), Ch. 5, 50–73.
51. Weidner, J.P., (ed.), *Wet-strength in Paper and Paperboard*, Tappi Monograph Series No. 29, Tappi Press, New York (1965).

52. Pakinkis, F., Developments in wet-end starch technology, in *World Pulp and Paper Technology 1990, Int. Review for the Pulp and Paper Industry*, ed. Roberts, F., Sterling Publications International Ltd, London (1989), 263, 266–269.
53. Allison, C.J., 'Testing Methods', in *Wet-strength in Paper and Paperboard*, ed. Weidner, J.P., Tappi Monograph Series No. 29, Tappi Press, New York (1965), Ch. 10, 131–143.
54. Kamutzki, W., 'Wet-strength resins of low formaldehyde content', *Papier* **41** (10A) (1987), 36–44.
55. Chan, L.L. and Martinez, E., in *Proc. 1989 Tappi Papermakers' Conference*, Tappi Press, Atlanta (1989), 357.
56. Elliott, R.G. and Drapeau, W.G., in *Proc. 1989 Tappi Papermakers' Conference*, Tappi Press, Atlanta (1989), 375.
57. Troemel, G. 'AOX-value reduced with new set-strength agents', in *World Pulp and Paper Technology 1990, Int. Review for the Pulp and Paper Industry*, ed. Roberts, F., Sterling Publications International Ltd., London (1989), 263, 266–269.
58. Stange, A.M.W., 'Wet-strength paper and additives in Europe', in *Wet-Strength Resins and Their Applications*, ed. Lock L. Chan, Tappi Press, Atlanta, Georgia (1994), Ch. 6, 101–108.
59. Pelzer, R., Kamutzki, W. and Moller, K., 'Wet-strength agents today – tomorrow?' *Wochenbl. Papierfabr.*, **117** (11–12) (1989), 499–500, 502, 504.

8 The sizing of paper with rosin and alum at acid pHs

J.M. GESS

8.1 Introduction

The sizing of cellulose fibres by rosin and its derivatives over the pH range 4.2–6.5 is discussed. The discussion includes the definition of sizing as both a surface treatment and a wet-end process. The various forms of rosin used in the sizing process are also described, as are the theories that have been presented to account for the interactions between rosin and cellulose fibres. Sizing is specifically described as a wet-end process.

8.2 Background

8.2.1 *Sizing*

The meaning of the term sizing depends on whether or not one is referring to the resistance of the sheet of paper or board to the sorption of water, or to the water-resistance of the constituent cellulose fibres. This distinction is extremely important, because the resistance of a given paper or board to the sorption of water is a function not only of the surface characteristics of the fibres, but it is also dependent on the structure of the web (i.e. the size of the pores, its density, the surface treatments to which it has been subjected, etc). It also depends on the properties of the test fluid being used to measure sizing and on the test procedure itself [1]. Because of this plurality of effects, it is extremely important to define which element of sizing is being discussed.

This chapter is concerned solely with one of the processes that leads to the development of water resistance in papermaking fibres, specifically the reaction of these fibres with rosin under acid conditions (up to a pH of 6.5). The terms papermaking fibres and cellulose fibres are used synonymously. The chapter will not deal with the topic of the resistance of paper and board to the sorption of water.

8.2.2 *History of rosin in papermaking*

The use of rosin as a means of rendering papermaking fibres resistant to the sorption of water is generally credited to Illig in 1804. However, there is a question as to whether the actual inventor was Illig, D'Arcet and Merimee or Carson. Discussions of the origins of rosin–alum sizing can be found elsewhere [2,3].

Rosin is used in the papermaking process in two forms: so-called neutral size (sodium rosinate), and the acid size (rosin acid). Traditionally, the neutral types of rosin size have tended to be used in papermaking systems where the calcium content, as defined by hardness of the water, was 50 ppm (as calcium oxide) or less. Papermakers found that, in hard water (where the calcium oxide content was greater than 50 ppm), the way in which the neutral rosin preparation was used was critical and there was little room for error. If this was not optimised, sizing efficiency dropped and the calcium rosinate formed by reaction of the neutral size with calcium ions in the system tended to precipitate in the lines and through the papermaking systems giving deposit-related problems. The acid size emulsion products were not as efficient as the neutral rosin sizes in soft water papermaking systems, but were easier to use under hard water conditions.

From the 1950s onwards, the rosin sizing process and also our understanding of it underwent many changes, for example, the development of fortified sizes, which are natural rosins modified by maleic anhydride or fumaric acid via the Diels–Alder reaction [4]. Microparticle size rosin acid emulsions were also developed [5]. The clarification of the work of Lorenz [6], Strachan [7], Wieger [8] amongst others by Strazdins [9], Davison [10] and others [11] and the work of Swanson [12,13] were also important developments during this time. New methods of monitoring sizing in paper via penetration measurements such as the Hercules size test [14] were also important developments and, most recently, the development of a new sizing theory to explain sizing in terms of differential bond strengths [15].

Each of these elements has played a major role in increasing the use of rosin in the papermaking process and in increasing the efficiency of the process.

8.2.3 The forms of rosin used in papermaking

8.2.3.1 *Neutral rosin.* Three forms of neutral rosin are currently available to papermakers:

(1) An 80% saponified product generally marketed as a 70–80% solids fluid. This product is normally manufactured by reacting rosin acid with sodium carbonate. It can be pumped at elevated temperatures, but requires special equipment to dilute it to the 5% concentration normally used in a papermill.

(2) A fully saponified product marketed as a 100% solids powder. This material is readily soluble in water. It is normally used by papermills located far from a size manufacturing facility, where the cost of shipping water is a major factor.

(3) A 50% solids containing system, which is known as a thin size, and which can be readily pumped and diluted with no special equipment. This material, at least in one form, consists of a 50–50 mix of neutral size and urea, and can be used in other combinations as described by Kulick [16].

In the past, it was known that the fully saponified product could not be readily handled at a consistency of greater than 50%. It was then found that a combination of 80% neutral rosin and 20% rosin acid could be shipped and used as 80% solids. Because both the neutral and acid rosin components were sizing agents, the economics were most favourable by shipping the high solids 80% saponified product which then became the primary neutral size marketed for many years. The thin sizes (50% solids) have been in use for a relatively few years.

8.2.3.2 *Rosin acids.* Rosin acid is available in a variety of emulsions. Historically, the most prominent of these was obtained by the Gillet [17] and Bewoid [18] processes. Only a small amount of Gillet process rosin acid size is marketed today, but there are still significant quantities of the Bewoid type emulsions being sold throughout the world. The original Bewoid type emulsions were stabilised by casein, a protein derived from milk, but many of the modern Bewoid type emulsions use soy protein as the stabiliser.

In 1974, Hercules developed and patented a means of manufacturing a microparticle rosin emulsion (0.1 micron as compared with 1.0 micron) that had no stabilisers. This replaced many of the rosin acid emulsion products used by papermakers [19]. Subsequently, other manufacturers developed the means of manufacturing microparticle emulsion products and these are the major rosin acid systems in use today [20]. The microparticle emulsions have a 35% solids content and, as with the thin size described earlier, are only suitable for transportation over short distances. The original Hercules microparticle size emulsions were primarily size stabilised and did not contain any chemical stabilisers. As a result, they were not shear stable. Until papermakers learned how to pump these materials, there were significant problems with precipitation in places other than onto the fibres.

8.2.3.3 *Modified rosin products.* It was found that the efficiency of rosin as a sizing agent could be markedly improved by reacting the levo-pimaric acid component with maleic anhydride via the Diels–Alder reaction (Figures 8.1 and 8.2). Subsequently, it was found that fumaric acid (and amongst others itaconic and other acids which would give a Diels–Alder reaction with the levo-pimaric form of rosin) would give materials which were at least as efficient as the maleic anhydride-modified rosin product [21,22]. This was an interesting development, because reacting rosin (which has one carboxyl group) with either maleic anhydride or fumaric acid gives a product which has

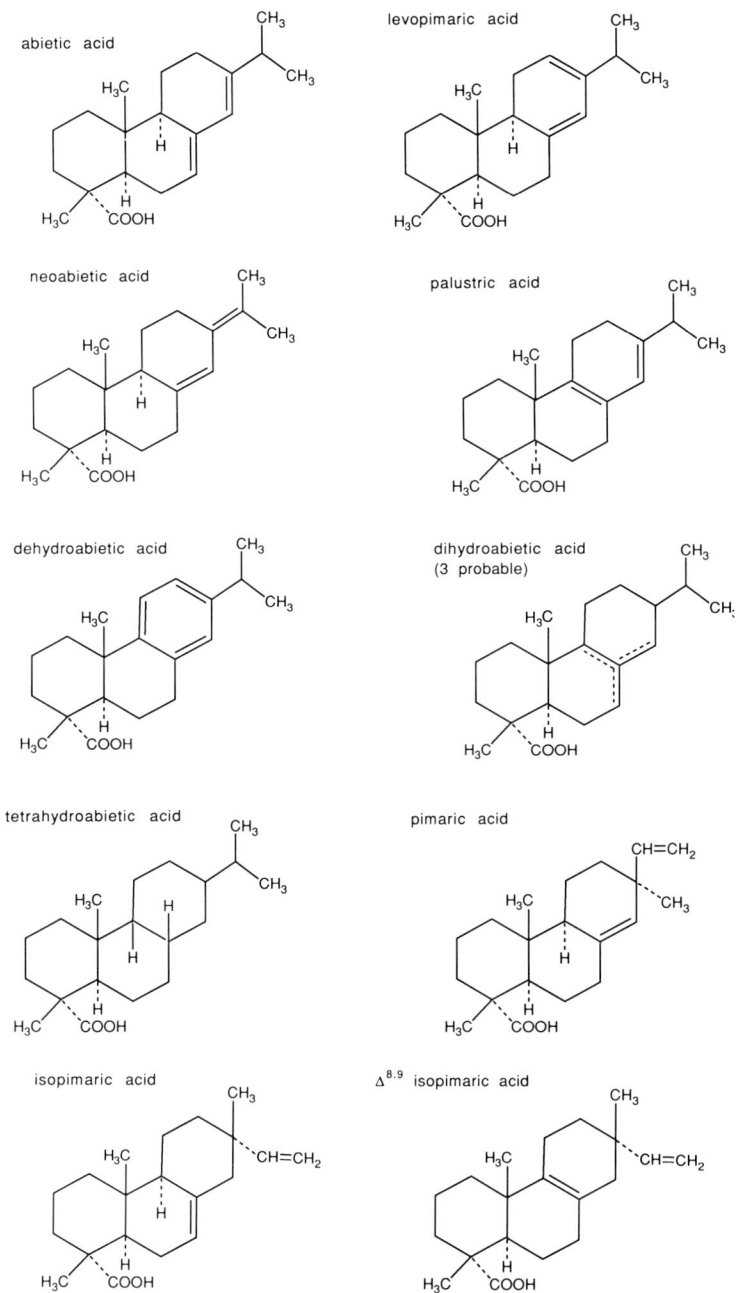

Figure 8.1 Forms of rosin in wood.

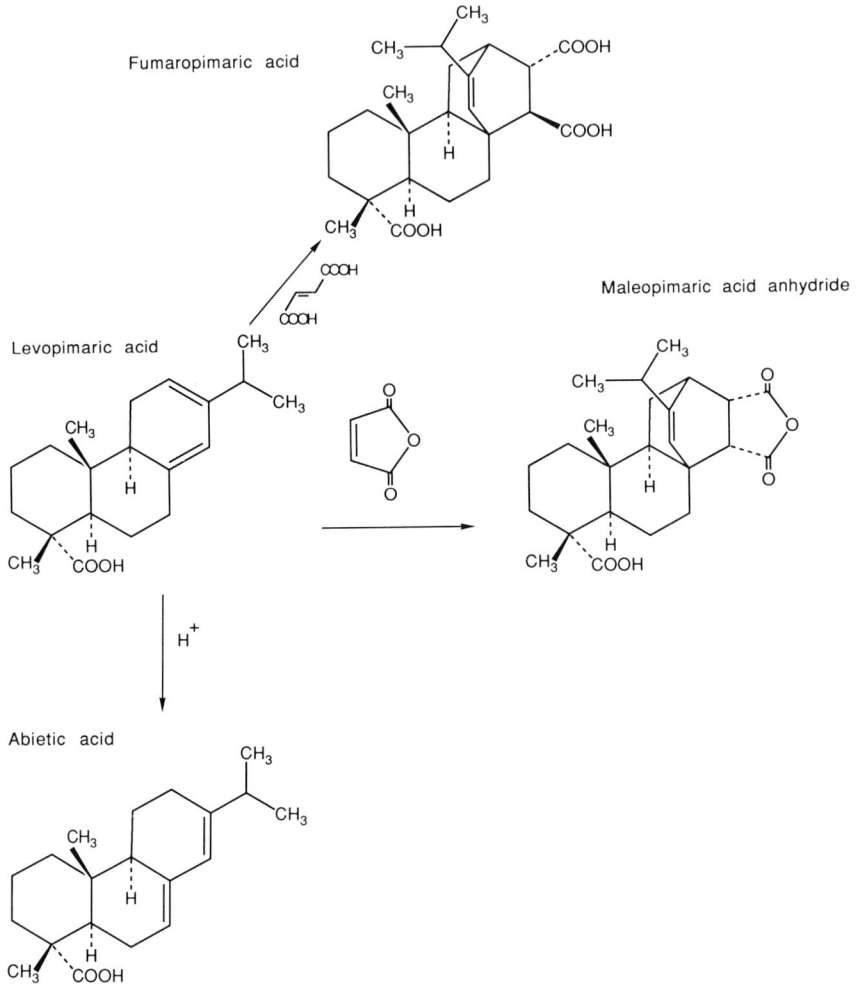

Figure 8.2   Primary reactions of rosin of interest to papermakers.

two additional carboxyl groups and is significantly more hydrophilic than the original starting material. Yet, the modified rosin is a better size. This may be due to the increased anionicity of the modified rosin product which thereby decreases the tendency of the size to agglomerate and increases the efficiency of the size by spreading it more uniformly through the fibres in the paper. This increase in efficiency more than balances the loss in hydrophobicity [23]. Alternatively, the increase in the anionicity of the size may increase the efficiency of its reactivity with alum to give aluminium rosinate (a material that is postulated to be the key to obtaining a sized product [23]).

There are no data to determine if either of the above is the predominant reason for the increased efficiency of Diels–Alder modified rosin. The tendency has been to assume that it is the increased distributional efficiency which accounts for the superiority of the Diels–Alder modified rosins. This is because it has been found empirically that the most efficient size is one that contains 20% of Diels–Alder modified product and 80% unmodified product.

It has also been found that the fumaric acid Diels–Alder reaction product with rosin gives a more stable product when used as an acid size in its emulsion form than does the maleic anhydride reaction product [5]. For this reason, the fumarate has become the more widely used material.

Recently, Mitsubishi [24] introduced a petrochemical derived size manufactured by the ene reaction (see also Chapter 9) between a long chain alkene, with a carbon chain the range of 22 carbon atoms, and maleic anhydride. The reaction product is hydrolysed using potassium hydroxide and is marketed in the soap form.

8.2.3.4 *Extended (thin) rosin sizes.* Rosin acid emulsions (acid size) were significantly easier to use than were the neutral rosin systems. The rosin acid emulsions could be delivered to a mill at 35% solids where they could be diluted to mill use consistency (5% solids). However, the price for the easy handling was the cost of shipping (65% water) and as a result, companies merchandising the rosin acid emulsions had to locate manufacturing plants near papermills. Neutral rosin size on the other hand could be delivered at between 70–80% solids and therefore could be shipped from a central manufacturing plant at a much reduced cost. This size however had a very high viscosity and required special handling techniques for pumping and dilution to the concentration required for use on a paper machine. This involved a significant cost to papermills.

In the 1970s, a new rosin product made its appearance, called extended rosin size, having a concentration of 50% [25]. The product could be diluted with water and had the efficiency of neutral rosin. Since that time, it was found that the use of potassium hydroxide instead of sodium carbonate or sodium hydroxide to neutralise the rosin acid, gave a product that matched the rosin/urea product in solids content and ease of dilution. These rosin sizes, the potassium soaps of rosin were collectively called thin sizes.

8.3 Sizing theories

8.3.1 *Structural considerations in sizing*

When sizing is discussed, it must be specified whether one is discussing the result of the measurement of sizing of a given sheet of paper or the sizing of component fibres. This is extremely important because, if one is determining the level of sizing in a given sheet of paper by one of the methods that measure the penetration of an aqueous fluid into a sheet of paper, one is measuring the balance of three factors, only one of which has to do with the actual sizing of the fibres themselves.

8.3.1.1 *Rate of penetration of the test fluid.* The rate at which the test fluid will penetrate the surface as compared with the rate at which it will penetrate the inner layers of the paper being tested must be considered. If the paper being tested has been passed through a size press, there will be a relatively dense layer of material on the surface of the sheet through which the test fluid will have to penetrate. Figure 8.3 shows that there can be a significant difference in the rate at which a test fluid will penetrate the surface layers of a web and the rate at which it penetrates the inner layers of the web. This effect can also be seen in Figure 8.4 where the penetration of a test fluid into a sheet of pulp was plotted against time. In this case, one surface of the sheet of pulp was heavily compacted by undergoing crushing on the wire side. This change in the density of some of the surface layers is clearly shown in the rate at which the test fluid penetrated the web.

8.3.1.2 *The size of the pores in the paper.* An aqueous test fluid can penetrate a sheet of paper either via the pores in the sheet (the areas between the fibres in the web) or by way of the fibres. The larger the average pore size is in a given sheet of paper, the greater is the probability that the fluid will penetrate the sheet via the pores rather than the fibres. Furthermore, the larger the pore size in a web, the greater is the possibility that the actual degree of the sizing of the fibres will have little influence on the rate of penetration of the fluid. This is interesting because, traditionally, it has been assumed that fibres manufactured from pine trees from the southern USA are harder to size than fibres made from more northern species of evergreen [26]. The reason for this assumption is that more rosin is needed in a sheet made from a southern USA

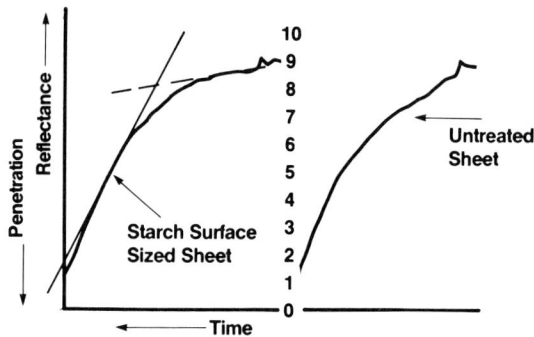

Figure 8.3 Rate of penetration of water into paper during the Hercules size test.

(a)

Addition of Water

Penetration

Reflectance

Time

(b)

Addition of Water

Penetration

Reflectance

Time

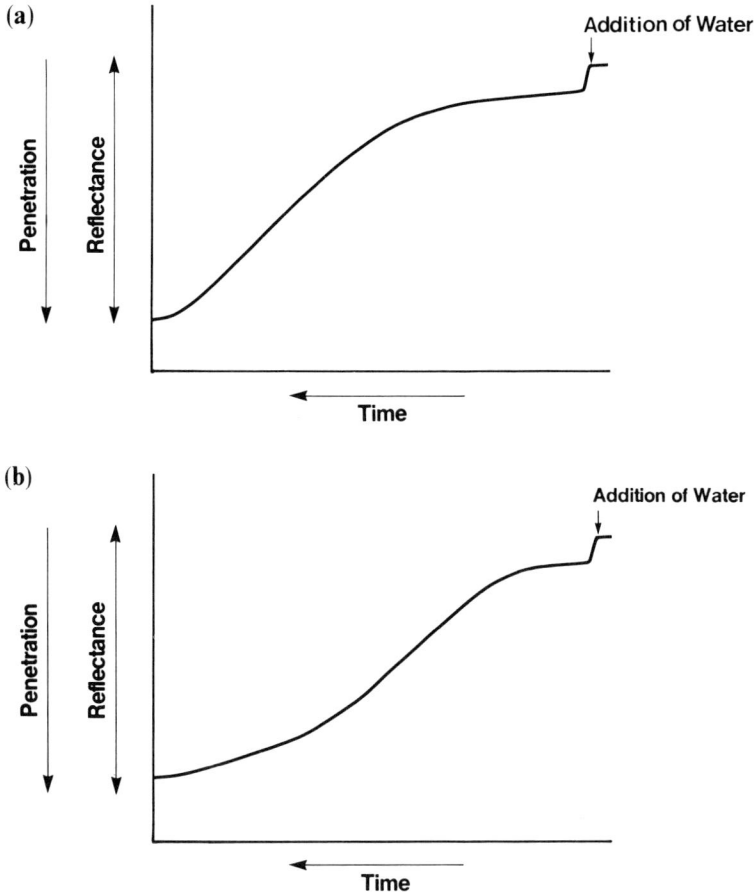

Figure 8.4   Rate of penetration of thickened water (with hydroxyethyl cellulose) through a pulp sheet. (a) wire to felt and (b) felt to wire.

pine than for a northern USA evergreen to give the same level of sizing as measured by a penetration test. The true reason, however, is that papers manufactured from the southern pines tend to have larger pores than do the northern USA papers. Thus, it is necessary to add more size to compensate for the larger pore size.

8.3.1.3   *The degree of hydrophobicity of the fibres in the sheet.* Fibre hydrophobicity only comes into play if the above two factors are negligible, and, in general, they are not. Therefore, although the papermaker may be treating the fibres to be formed into a sheet with sufficient material to render them completely hydrophobic, it is still possible to produce a sheet which has a low level of sizing as measured by a penetration type test. It is for these reasons that the various tests used to give a measure of the sizing in a sheet of paper give results which tend not to correlate.

As an example, the sizing measurement test based on contact angle tends to measure the hydrophobicity of the surface layers of the paper being tested. If the fibres in the paper are highly sized, but the paper has an open structure and there has been no surface treatment to cover the sized fibres, then the web will show a high contact angle. If the same web is tested by a penetration type test, the sizing level will be low. It is also for this reason that, in any discussion of sizing, the effect of such elements as sheet structure on the measurement of sizing should always be considered.

8.3.2 *The interaction between rosin and cellulose fibres leading to sizing*

8.3.2.1 *Historical developments.* The development of the understanding of how rosin and alum work together to impart sizing to papermaking fibres falls into three periods. Firstly, the period beginning with Illig and continuing into the 1920s, which was concerned primarily with understanding what were the products of the reaction of anionic rosin with alum and anionic cellulose fibres [2]. Secondly, the period beginning in the 1950s, which was primarily concerned with how the flocs of rosin, which were precipitated onto the cellulose fibres, could migrate and be uniformly distributed through the mass of fibres in a sheet of paper to give sizing. The work in this period also laid down the criteria necessary for sizing agents to impart sizing to cellulose fibres and paper [27, 28]. Thirdly, the period beginning recently, which has been concerned with improving our understanding of how rosin is oriented onto cellulose fibre surfaces in order to maximise its potential to impart hydrophob-icity and, at the same time, how it is able to move through a sheet of paper and achieve a distribution necessary for good sizing [15, 29].

Much of the early work from Illig to Ostwald and Lorenz has been well summarised by Holtz *et al.* [30], Sutermiester [2], Strachan [7] and others [31]. Much of what is covered here is taken from their work.

In his original work, Illig (see reference [2]) postulated that sizing resulted from the precipitation of free rosin, by the action of acidic alum, on to cellulosic fibres. This was expanded by others who pointed out that permanent sizing did not result unless there was an agent such as alum present in the system [7]. In other words, more than just the acidity of the alum was required for permanent sizing to occur. As in the reaction of acid dyes with cotton fibres, a mordanting agent was required. From this work, a new hypothesis was developed where it was postulated that the active agent in the sizing of cellulose fibres was an aluminium–rosin complex often referred to as aluminium resinate (or rosinate). Further work during this period showed that one result of the addition of alum to cellulose fibres was to reverse the charge on the fibres from negative to positive. From this work developed the electrostatic theory of sizing which explained how it was possible for anionic rosin size emulsions to be selectively precipitated on to cellulose fibres [32].

Using this work as a base, Strazdins [32], Davison [23] and others [11] laid down what could be considered to be the rules that were required for any sizing agent, including rosin in any form, to impart sizing to the cellulose fibres. From these rules and from IR spectral data of the reaction products of both rosin acid and sodium rosinate with alum under acid conditions, the so-called flow or sintering theory of sizing developed.

Metzler [33] showed that, although the IR spectra of the reaction products of rosin acid and sodium rosinate with alum were similar, it was possible to differentiate the products by solvent extraction. Free rosin acid could be extracted from the product of reaction of rosin acid with alum by the use of an appropriate solvent. It was not possible, however, to separate rosin from alum when the product was obtained from the reaction of sodium rosinate with alum (i.e. aluminium rosinate).

With Metzler's work as a base, the third period began in which our understanding of how rosin and alum impart permanent sizing to cellulose fibres developed. This has resulted in the development of a new theory of sizing [15, 29].

8.3.2.2 *General considerations.* It has been postulated and generally accepted that, for rosin or any other sizing agent to be effective, the following conditions (based on the premise that a continuous film is not formed on the surface of cellulose fibres) have to be met [30]:

- The size must be oriented on the cellulose fibre in order to optimise its hydrophobicity (Figure 8.5). That is, the hydrophilic head of the rosin must be oriented towards the fibre and the hydrophobic tail must be oriented outwards.
- The size must be evenly distributed through the paper.

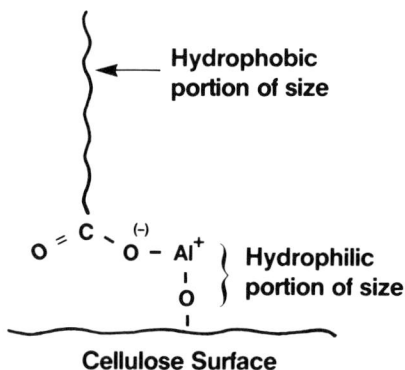

Figure 8.5   Orientation of size molecule on cellulose surface to optimise sizing.

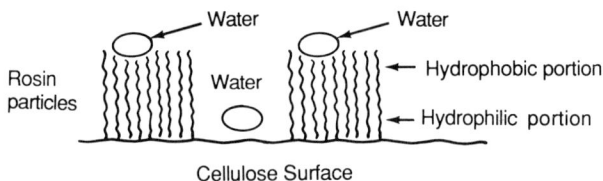

Figure 8.6 Spacing of rosin on cellulose to give sizing.

- The spacing between the size molecules on the fibres must be such that water is effectively repelled (Figure 8.6).

In this context it is relevant to note that if the surface of the paper is sealed by a coating or a continuous film, then it will exhibit sizing by most test procedures except those measuring wicking.

Rosin is most often used with alum at a pH of 4.2–5.5 to size the fibres in paper. It is possible to use a low molecular weight, high charge density polymer with the rosin instead of alum, but this tends to be inefficient.

8.3.2.3 *The flow or sintering theory of sizing [23, 32].* The theory assumes that, in a rosin–alum system, rosin in either its free acid or soap form will react with the aluminium moiety of alum to give the aluminium ester product. In the case of the reaction of neutral rosin (sodium rosinate) with alum under acid conditions, most of the rosin reacts rapidly to give the aluminium ester. The rosin acid that is formed reacts with the remaining aluminium in the system in the drier section of the paper machine.

It is the reaction product of the rosin and aluminium moiety of the alum that provides the anchoring of the rosin onto the surface of the fibre in an orientation which maximises sizing. Only one product is postulated – the aluminium ester. A question remains as to how the aluminium ester of the rosin is anchored onto the cellulose papermaking fibre. This may be by electrostatic attraction or by the formation of an aluminium ester bridge.

The reaction product of neutral rosin and alum has a positive surface charge which makes it substantive to cellulose. By virtue of its structure, the positively charged product is oriented onto the cellulose in a manner which optimises its hydrophobicity.

The rosin acid that is precipitated onto the fibre, either from the breaking of a rosin acid emulsion or by its formation as a by-product of the neutral rosin/alum reaction, migrates through the sheet via a flowing or sintering mechanism. It eventually reacts with alum to give a product analogous to that which is formed from the reaction of sodium rosinate and the aluminium moiety of alum.

This hypothesis, developed by Strazdins [9], Davison [23] and others [11], answered many of the questions that had been raised by papermakers over the years but, as the use of rosin acid emulsions increased, questions were raised as to its viability, particularly whether it was general enough to explain the reaction of all sizing materials or if it was only specific to the reaction of rosin soap and alum.

8.3.2.4 *Anomalous data.* Anomalies began to appear which were not accounted for by this flow theory. In particular, Metzler [33] showed that it was possible to separate aluminium rosinate into a series of species by the use of solvents. In particular, it was possible to separate the rosin from the aluminium under certain conditions. Based on this work, it was suggested that more than one reaction product was possible. These fell into two categories: the aluminium rosinate ester which could not be separated by solvents, and aluminium rosinate complexes which could be separated by solvents. Metzler's work was interesting because he found that both the aluminium rosinate esters and the aluminium rosinate complexes gave approximately the same IR spectra. This observation was the basis of the work which finally postulated that the end product of the sizing reaction was aluminium rosinate.

Further work by Metzler [33] showed the presence of rosin–aluminium complexes in the sizing reaction. Rosin acid and alum were mixed and heated overnight at 105°C. Almost all the rosin could be extracted from the rosin–alum mix, indicating that little or no aluminium rosinate had been formed.

A series of studies by Swanson [12, 13] showed that it was possible to size paper by the vapour phase migration of stearic acid through paper. No alum was necessary for sizing in this way.

Gess [34] showed that rosin acid was present in the hood of a paper machine under conditions where the web was sized with neutral rosin and alum at high total acidities (i.e. under conditions where a significant percentage of rosin acid would be formed). The web drained poorly and was wet coming into the drier section of the paper machine. The first cylinders in the drier section were heated to a maximum to maintain production.

Studies by Neogi and Jenson [35] showed that the sizing in a roll of linerboard (thirty-eight layers down) was significantly less than the level of sizing in the surface layer of the roll. There was a migration of size from the inner to the outer layers of the roll. The paper had been manufactured with a combination of rosin acid and alum. The migration occurred when the rolls were wound hot at temperatures as high as 160°F. Major [36] showed that the migration of sizing in a roll could be minimised by the addition of neutral rosin (sodium rosinate). It is fascinating that the neutral rosin used in this work came into the system by admixing pine with the Douglas fir pine prior to pulping.

Sizing regression, i.e. the diminution of sizing in paper with time, had been observed and was assumed to be due to a deficiency of aluminium ions. This arose either through a decrease in the amount of alum used or from an increase in the amount of competing anionic carry-over which, in turn, used up some or all of the available alum. The problem of regression was solved by increasing the addition of alum.

Cationic wet-strength resins were found to increase sizing efficiency. Therefore, it seemed that cationic elements other than those provided from alum could give sizing. The strong bond/weak bond theory of sizing (see section 8.3.2.5) suggests that the fixing of dyes and sizing are parallel processes. This is especially so in that alum and certain cationic polymers can be used as dye fixatives.

Gess [34] showed that sizing from a sheet of paper treated with rosin acid and alum could migrate through four layers of alum-treated paper in a 21-day period when the papers were heated at 105°C. Gess and Tiedeman [29] showed that the rate of this migration could be related to factors such as the vapour pressure of the size, the bond energy between the size and the substrate (i.e. the bond energy between the molecule that bridges the size and cellulose substrate) and the number of sorption/desorption steps required for distribution of the size through the web to give hydrophobicity.

A further anomaly is that demonstrated in experiments carried out by Marton [11] which showed that the mechanism of the reaction of neutral rosin and alum was different from the mechanism of the reaction of rosin acid and alum.

8.3.2.5 *The strong bond/weak bond theory of sizing [15]*. In studying these anomalies, it seemed that the sintering or flow theory of sizing was most probably limited only to sizing via reaction of neutral rosin with alum under acid pH conditions to give aluminium rosinate ester. The theory did not explain the many anomalies and it was therefore necessary to extend it. The strong bond/weak bond mechanism [15] is an attempt to bring together all the facts and data known today. The theory assumes that a sizing system contains four components: the sizing agent, the precipitator (the material that brings the size down onto the fibre), the fixing agent (the material that binds the size to the cellulose) and the substrate (cellulose). Under specific conditions, one component can have more than one function. For example, alum between pH 4.2–4.8 is both a precipitator and a fixing agent. At pHs in the range 5.5–6.5, alum retains its precipitator function, but tends to lose its function as a fixing agent [15].

There is more than one type of bond that can be formed between the sizing agent, the fixing agent and the substrate. These bonds can be broadly divided into two groupings: strong bond systems, where the bond strength between the three elements is such that the size cannot migrate under the energy conditions available in the drier section of the paper machine, and weak bond systems, where the bond between the elements in the papermaking system is such that the energy available in the drier section of a paper machine is able to break the bonds in the system and allows the sizing agent to migrate.

Experience indicates that the aluminium rosinate ester formed by the reaction of sodium rosinate and the aluminium moiety of alum at acid pHs forms a strong bond complex with cellulose and does not migrate. The complex of rosin acid and alum forms a weak bond complex which can be broken and which allows the rosin acid to migrate.

8.3.3 *Implications of the strong bond/weak bond theory*

Based on the strong bond/weak bond concept, it is postulated [29] that the development of sizing in a paper depends on the bond energy between the sizing agent, fixing agent and substrate, the vapour pressure of the sizing agent and the number of migration steps required for the sizing agent to spread through the web to give uniform coverage. It also depends on the energy of the bond that is formed between the rosin, the fixing agent and the substrate.

The size and distribution of the rosin on the fibres when it is precipitated is also important. If it is precipitated in large flocs, then the heating of the rosin-treated fibres (in the form of a sheet) will not give sufficient lateral movement for coalescence to take place. An important consideration in this context is that, during the heating of the sheet in the drier section of a paper machine, the major movement of the size particles is basically away from the heat [35]. It is most probable that lateral movement of the rosin particles is very limited.

It is therefore possible to conclude that, in a system where a strong bond complex is formed (i.e. by reaction between neutral rosin soap and the aluminium moiety of alum to give aluminium rosinate), it is necessary for the size to be uniformly distributed through the mass of the papermaking fibres in the system before the addition of the alum. If this does not occur, then the size will not be uniformly distributed onto the fibres and the result will be poor sizing.

In a system where the primary reaction product is a weak bond complex, then the initial distribution of the size, although important, is not critical. For it is assumed in this theory that migration occurs via the vapour phase and, therefore, although there is some lateral movement of the size, the primary direction of the movement of the size through the web is perpendicular [36]. Therefore, if the size is present in large flocs, migration will not cause the surfaces of the fibres to be covered uniformly. It is for this reason that the microparticle size emulsions, having a particle size around 0.1 micron, are more efficient than the 1–3 micron Bewoid size systems.

It is now possible to explain why, in a neutral rosin/alum system, one level of sizing is obtained when the sheet is dried under ambient conditions (TAPPI Standard Conditions) and, when the sheet is heated, an incremental gain in sizing occurs. The initial sizing level is due to the reaction giving aluminium rosinate as a product (i.e. the strong bond complex product). The incremental gain when the sheet is heated is due to the migration of that portion of the size precipitated as the rosin acid (i.e. that portion of the size that will react to form a weak bond complex with the aluminium moiety of the alum).

When fibres are treated with a rosin acid emulsion and alum, and formed into a sheet of paper and dried at ambient conditions there is little sizing. With heating, sizing develops. When fibres that have been pretreated with a cationic polymer are treated with rosin acid emulsion and alum, some sizing is found when the sheet is tested after drying under ambient conditions. Again, a significant increase in sizing takes place when the sheet is heated [36].

There is always some migration of size during drying (because of the vapour pressure of the size). When alum is the fixing agent, the rosin–alum complex bond, although weak, is still strong enough to minimise migration when the sheet is dried at ambient conditions. Where there is a rosin acid–cationic polymer bond, there is migration because this bond is extremely weak [15]. This relationship between bond strength and rate of migration was shown most graphically by Tiedeman [29] who developed one material which, in 24 hours, could migrate through four sheets of paper treated only with a cationic polymer as a fixing agent. When the study was repeated using the aluminium moiety of alum as the fixing agent, the size was retained in the top sheet of the stack [34].

The observation, in sizing regression studies [37], that the addition of more fixing agent decreases the potential of the sheet to show sizing regression, can now be explained. It is suggested that, initially, the rosin forms a hydrogen bonded complex with the alum, as shown in Figure 8.7(A). However, as both the substrate and size tend to have anionic charge, the bond is extremely weak and, in time, the size will rotate to a more stable configuration as shown in Figure 8.7(B). In this configuration, with the hydrophilic head oriented outwards, the sizing agent does not impart hydrophobicity to the fibre substrate.

It is interesting to note that this theory of sizing supports the view, proposed by earlier workers [38], that there is a relationship between sizing and the direct dyeing of papermaking fibres. Direct dyes tend to be substantive to

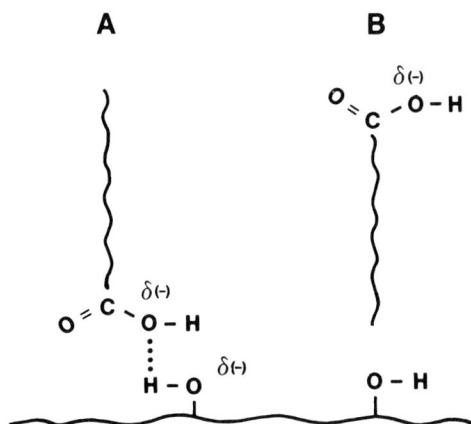

Figure 8.7 Mechanism of the regression of sizing by inversion.

papermaking fibres, but the efficiency of their retention can be augmented by the addition of alum or a low molecular weight, high cationic charge density polymer.

The discussion in this chapter deals solely with rosin and its derivatives. There is no discussion relating to alum and its various structures and forms as seen in other treatises on rosin/alum sizing. This is deliberate because the strong bond/weak bond theory of sizing is only concerned with the energy of the bond between the size, fixing agent and substrate. The actual form of the alum in the end is less important to the papermaker than knowing the potential of a given size system to migrate in the drier section of a paper machine. For a study of alum chemistry, a more detailed treatment is available elsewhere [39].

8.4   Rosin sizing at pHs greater than 5.5

Rosin in one form or another has been used in the sizing of paper at pHs in the range of 4.2–5.5. With alum alone as the precipitant and orienting agent, sizing falls off rapidly as is shown in Figure 8.8 [40]. However, as far back as 1932 [41], there was interest in sizing paper at pHs above 4.5. Curtis, in his patent claimed that papers could be sized at pHs in the range of 6.5, by treating fibres with rosin (40 lb/ton), followed by alum, followed in turn by sodium aluminate. The patent did not specify the form of rosin that was used, but sizing could be obtained in the pH range 5–7.

Libby [47], in 1933 obtained a patent for sizing paper with rosin at pHs above 5.5, but used only sodium aluminate. This work was reproduced by Chene in France in 1961 [43].

In 1967, Davidson [44] studied the sizing of paper containing whiting (chalk) with rosin and alum and noted that sodium rosinate was prone to

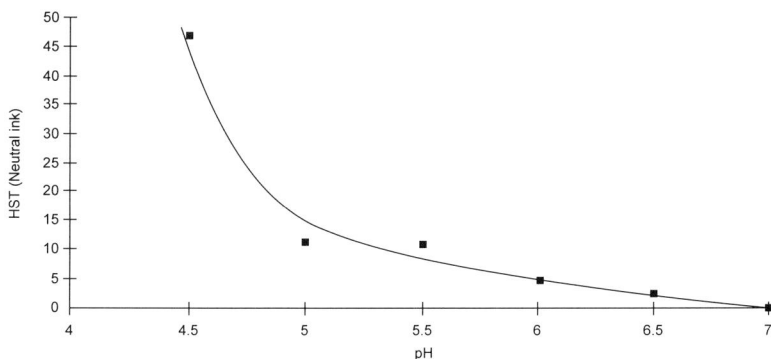

Figure 8.8   Sizing *vs.* pH (rosin acid, alum).

react with the dissolved calcium in the system (from the chalk) leading to the formation of a stable calcium rosinate foam which presented operating problems. Dispersed rosin size (rosin acid emulsion size) gave sizing and also fewer operating problems than the sodium rosinate size. Davidson therefore concluded that if rosin size was used, it was possible to size paper with rosin and alum at the pHs when whiting (chalk) was the pigment of choice (i.e.; pHs above 6.0). This work is important because it was one of the first attempts to study sizing where a form of calcium carbonate was used in the furnish. Davidson's observations concerning the effect of calcium ions was also shown by Sturmer and Poppel [45].

Gess [40] showed that the combination of a high charge density, low molecular weight cationic polymer and alum gave good sizing efficiency (Figure 8.9). In this study, the cationic polymer was added to the fibres prior to the addition of the rosin acid emulsion, and this was followed by the addition of alum. However, as is shown in Figure 8.9, although there is a good sizing efficiency, it reaches a maximum at a pH of 6.0 and decreases steadily above this pH. However, a significant level of sizing still remains in the sheet at 7.0.

Peck and Markillie (46) showed that sizing efficiency could be improved by pre-mixing the alum with the size emulsion. However, they also observed that sizing efficiency maximises at a pH around 6.0.

Wortley [47] showed that substituting polyaluminium chloride (PAC) for alum increased the efficiency of a rosin/aluminium containing system at pHs above 5.5. Colasurdo [48], and Liu [49] also studied the sizing of paper at near neutral pH levels substituting polyaluminium chloride for alum. Polyaluminium chloride maintains its positive character to a higher pH than does aluminium ion and is thus less prone to react with the hydroxyl ions in the water. Polyaluminium chloride simply increases the potential for the reaction of rosin with aluminium portion of the reaction system.

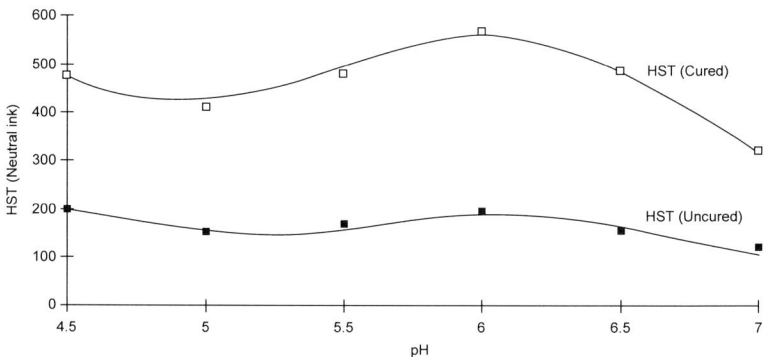

Figure 8.9 Sizing *vs*. pH (rosin acid, alum and a cationic polymer).

However, Gess [40] and others showed that alum could be used at pHs between 6.5 and 7.0 if it was added properly (i.e.; at a point where the reaction with the rosin could be optimised). Hegenbart *et al.* [50] and others showed that the use of polyethyleneimine (a low molecular weight, high charge density cationic polymer) in addition to polyaluminium chloride improved sizing efficiency. Earhart [51], reported that the efficiency of a neutral rosin system could be enhanced by the use of a cationic rosin emulsion.

The work described above, showed that it was indeed possible to size papermaking fibres with rosin at pH levels above 4.5. Efficiency was apparently enhanced at the 6.5–7.0 level by substituting polyaluminium chloride for alum and by the use of cationic polymers of the high charge density, low molecular weight type. Cationic rosin emulsions were another alternative.

Bierman [52–54], in a series of papers showed that it was possible to obtain sizing with rosin soap and ions other than aluminium. His studies are interesting in that he shows that the measurement of sizing by a penetration test using a formic acid containing test fluid gave poor results whilst a test fluid without formic acid (containing only dye and water) gave extremely good sizing. Gess [40] studied the rate of penetration of various test fluids into paper and showed that test fluids containing chelating agents (lactic acid, etc.) penetrated paper at a significantly faster rate than did a test fluid containing only dye and water. The change in penetration rate could not be explained by postulating acidity or surface tension effects.

It may be that the rosin in Bierman's studies reacted to form a weak bond type complex (see section 8.3.2.5) rather than a strongly bound product with the rosin. This type of complex may then be readily broken by a test fluid containing formic acid (the standard HST test fluid).

Bierman's work is interesting in that he was able to obtain sizing at pH levels in the range of 9.0. All of the previous work (with rosin/alum or rosin/polyaluminium chloride systems) could not achieve sizing above 7.5. Bierman [52] does present a hypothesis to account for these results based on dye mordant chemistry and on complex formation, but at this time his work has not been corroborated by others.

8.5 Summary

The various types of rosin available to papermakers have been discussed, as has the development of our understanding of how they impart sizing to cellulose fibres.

Rosin, in combination with the aluminium moiety of alum, can render a fibre resistant to the sorption of water over the pH range 4.2–5.2. It is proposed that, in the reaction of the rosin with the fibre, a discontinuous covering of the fibre is formed, with the molecules of rosin close enough to repel droplets of water from the fibre. For sizing to occur, the distribution of the rosin through the fibres must be relatively uniform.

A controversy exists as to how the rosin distributes itself through the papermaking system. The traditional approach is based on a melting and a physical flow of the rosin through the sheet of paper. A more recent hypothesis, the strong bond/weak bond theory, suggests that the rosin is distributed through the fibre mass via the vapour phase. By postulating a vapour phase migration, this theory has been able to reconcile the work of Swanson with that of Strazdins and others.

There is also controversy as to the bonds that can be formed between rosin, the fixing agent and the substrate. The traditional theory proposes that, under all circumstances, rosin will form an ester bond with the aluminium moiety of alum. This provides for the bond between the rosin and the cellulose substrate. The strong bond/weak bond theory postulates that there is more than one bond possible between rosin, the aluminium moiety of alum and the cellulose substrate. Further, these bonds can be differentiated by the amounts of energy required to break them and by their response to various solvents.

This new theory provides a means of rationalising the work of Marton who observed that rosin acid and neutral rosin work by two different mechanisms. The strong bond/weak bond theory now proposes one general mechanism which can account for how the two types of size work.

Finally, the term sizing is ambiguous in that it depends in normal use on how sizing is to be measured and on the structure of the sheet rather than on those reactions that lead to fibre hydrophobicity.

References

1. Gess, J.M., *Tappi*, **64** (1) (1981), 35.
2. Sutermiester, E., *Paper Trade J.*, **97** (4) (July 27, 1933), 18; **97** (21) (Nov. 23, 1933), 26; **97** (22) (Nov. 30, 1933), 20; **98** (1) (Jan. 4, 1934), 25.
3. Olsen, S.R. and Gortner, R.A., Technical Assoc. Section, *Paper Trade J.* (October 18, 1928), 153.
4. Wilson, W.S. and Bump, A.H., US Patent 2,628,981 (1957); Maor, A.C. Van der, and Wilhelmi, W.A., Canadian Patent 595,416 (March 29, 1960).
5. Aldrich, P.H., US Patent 3,817,768 (June 18, 1974).
6. Kumler, R., *The Papermill* (June 3, 1933), 8.
7. Strachan, J., *The Papermaker, British Paper Trade Journal*, **82** (1) (October 1, 1931), 36.
8. Wieger, B., *Zellstoff Papier*, **12** (1) (Jan. 1932), 15.
9. Strazdins, E., *Tappi*, **55** (12) (1972), 691.
10. Davison, R.W., *Tappi*, **47** (10) (1964), 609.
11. Marton, J. and Marton, T., *Pulp and Paper Magazine Canada*, **83** (11) (1982) 26.
12. Swanson, J.W. and Cordingly, S., *Tappi*, **42** (10) (1959), 812.
13. Swanson, R., *Tappi*, **81** (7) (1978), 77.
14. Price, D., Osborn, R.H. and Davis, J.W., *Tappi*, **36** (1) (1953), 42.
15. Gess, J.M., *Tappi*, **72** (7) (1989), 77.
16. Emerson, R., US Patent 34,606,791 (August 19, 1986).
17. French Patent 752,970 (October 4, 1933).

18. Wieger, B., British Patent 335,902 (1931).
19. Monsanto Bulletin on Monsize (undated).
20. Westvaco, personal communication.
21. French Patent 1,301,517 (July 9, 1962).
22. Japanese Patent 4155 (May 27, 1959).
23. Davison, R.W., *Tappi*, **58** (3) (1975), 48.
24. Takahashi, Y. and Yoshida, H., British Patent 2,317,614A (October 10, 1984).
25. Emerson, R.W. and Martin, J.L., U.S. Patent 4141750.
26. Vanderberg, E.J. and Spurlin, H.M., *Tappi*, **50** (5) (1964), 209.
27. Arneson, T.L., *Interaction of Alum and Cellulosic Fibres*, IPC Dissertation (1980).
28. Dennett, F. and Libby, C.E., *Tech. Assoc. Papers, Series XXIII* (1984), p. 617.
29. Gess, J.M. and Tiedeman, G.T., US Patent 4,587,149 (August 5, 1989).
30. Holtz (see reference 3).
31. Harrison, H.A., *World Paper Trade Review* (October 21, 1931), 1274.
32. Strazdins, E., *Tappi*, **60** (10) (1977), 102.
33. Metzler, R., personal communications.
34. Gess, J.M., *TAPPI 1982 Papermakers' Conference Proceedings*, TAPPI Press, Atlanta, GA, p. 9.
35. Neogi, A. and Jenson, J.R., *TAPPI 1984 Papermakers' Conference Proceedings*, TAPPI Press, Atlanta, GA, p. 179.
36. Major, H., personal communications.
37. Gess, J.M., unpublished data.
38. Sutermiester, E., *American Dyestuffs Reporter*, (August 11, 1924), 513.
39. Hayden, P.L. and Rubin, A.J., eds., *Aqueous Environmental Chemistry of Metals*, Chapter 9, Ann Arbor Science Publishers, Michigan (1976).
40. Gess, J.M., *Weyerhaeuser Research Report* TR-432, dated 12/16/83.
41. Curtis, F.J., US Patent 1,885,185; Nov. 1, 1932.
42. Libby, C.E., US Patent 1,929,205; Oct. 3, 1933.
43. Chene, M., *Le Papeterie*.
44. Davidson, R.R., *Paper technology*, **8** (4) (1967), 370.
45. Strurmer, L. and Poppel, G., *Wochbl. Papierfab.* **111** (19) (1983), 713.
46. Peck, M.C. and Markillie, M.A., *TAPPI 1994 Papermakers' Conference Preprints*, TAPPI Press, Atlanta, GA, p. 165 (Vol. 1).
47. Wortley, B., *Pulp and Paper*, **64** (11) (Nov. 1991), 131.
48. Colasurdo, A., *Preprints, TAPPI 1990 Neutral/Alkaline Papermaking Short Course* (Orlando, Florida), TAPPI Press, Atlanta, GA, pp. 71–78.
49. Liu, J., *Paper Technology*, **34** (8) (1993), 20–22, 24–25.
50. Hagenbart, K. *et al.*, *Wochbl. Papierfab.* **115** (7) (1987), 293–294, 296–297.
51. Ehrhardt, S.M., *Preprints, TAPPI 1990 Papermakers' Conference*, TAPPI Press, Atlanta, GA, p. 171.
52. Subrahmanyam, S. and Biermann, C.J., *TAPPI*, **75** (3) (1992), 223.
53. Biermann, C.J., *TAPPI*, **75** (5) (1992), 166.
54. Zhuang, J. and Biermann, C.J., *TAPPI*, **76** (12) (1993), 141.

9   Neutral and alkaline sizing

J.C. ROBERTS

9.1   Introduction

Internal sizing is achieved by retarding the rate of penetration of a fluid, usually water, through capillaries formed both within and between fibres. The rate of capillary rise of fluids into paper has been effectively described by the Washburn equation [1] and modifications of it [2]. These approaches assume that the penetration of a fluid into paper is analogous to that of a fluid into a single capillary. Retardation is thus brought about by the creation of a low energy hydrophobic surface at the fibre–water interface which increases the contact angle formed between a drop of liquid and the surface and thus decreases the wettability. Contact angles have been shown to be sensitive to molecular packing, surface morphology and chemical constitution [3–5], and it is the latter which is influenced during the internal sizing of paper. An excellent review is available [6] and this subject is also discussed more fully in Chapter 8.

The introduction of hydrophobic groups onto the fibre surface has to be done in a way which will allow a good distribution of the sizing molecule through the body of the sheet but which will not interfere with interfibre bonding (cf. fibre softening agents). Thus internal sizes are introduced at the wet end of the papermaking system, usually as colloidal suspensions, which are retained in the fibre network during sheet formation. During drying, and to some extent afterwards, they are able to migrate and undergo some form of interaction with the surface which allows orientation of the hydrophobic group and the creation of a low-energy surface.

In the acidic rosin–alum sizing process (see Chapter 8), the hydrophobe is a naturally occurring rosin acid which is retained by a primarily electrostatic mechanism [7]. The orientation of the size molecule takes place through the intermediacy of aluminium sulphate. No covalent bonding is believed to be involved. High pHs lead to the size being either poorly retained or retained in an inappropriately oriented form and thus to complete loss of sizing. The effect is clearly demonstrated by the series of photographs in Figure 9.1 showing the effect of a droplet of water on rosin–alum sized paper at pHs in the range 4.4–8.5.

Papermakers have, for reasons which are discussed in section 9.2, wanted to run at neutral or slightly alkaline pH and it is in response to this need that reactive sizes were developed. Unlike the rosin–alum system, these depend

pH 4·4   pH 4·8   pH 6·4

pH 7·4   pH 8·5

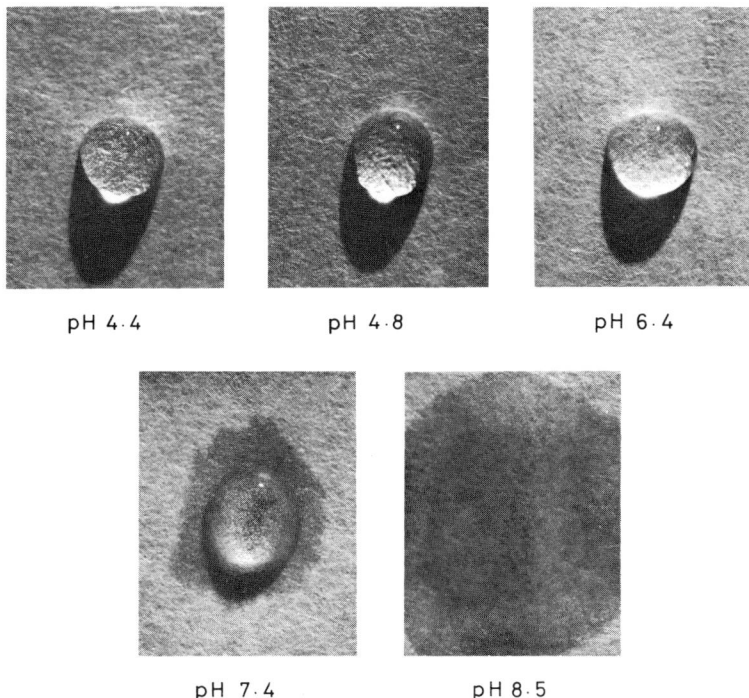

Figure 9.1   Droplets of water on rosin–alum sized paper prepared at different pHs.

Figure 9.2   General reaction involved in neutral sizing. R = hydrophobic group. X = reactive group.

upon the size molecule undergoing covalent attachment to the hydroxyl groups of the fibre surface through reactions such as esterification. A general reaction scheme is shown in Figure 9.2 and a list of the most common compounds which have been considered, but not always used commercially, is shown in Table 9.1.

The covalent linkage allows the permanent attachment of hydrophobic groups in a highly oriented state which makes sizing possible at very low levels. The main requirement of the molecule is that it has good stability towards water, since it is necessary to prepare it as a stabilised aqueous emulsion for use at the wet end, and that it also has good reactivity towards cellulose. These are, to some extent, mutually exclusive, and a compromise is therefore sought. A further discussion of this point is taken up in sections 9.3.3 and 9.5. In addition, the selected molecule must have physical properties which allow diffusion and

Table 9.1 Reactive sizes (commercial and theoretical) for cellulose

Compound	Structure	R	Cellulose derivative	Reference
Acid chloride	$R-\overset{\overset{\textstyle O}{\|}}{C}-Cl$	C_{14}-C_{18}	$R-\overset{\overset{\textstyle O}{\|}}{C}-O-Cell$	[8, 9]
Acid anhydride	$R-\overset{\overset{\textstyle O}{\|}}{C}-O-\overset{\overset{\textstyle O}{\|}}{C}-R$	$C_{17}H_{35}$	$R-\overset{\overset{\textstyle O}{\|}}{C}-O-Cell$ (+ RCOOH)	[10–12]
Enol ester	$R-\overset{\overset{\textstyle O}{\|}}{C}-O-\underset{\underset{\textstyle CH_3}{\|}}{C}=CH_2$	$C_{17}H_{35}$	$R-\overset{\overset{\textstyle O}{\|}}{C}-O-Cell$ (+ (CH$_3$)$_2$CO)	[13]
Alkyl ketene dimer	(see structure)	C_{14}-C_{18}	$R-CH_2-\overset{\overset{\textstyle O}{\|}}{C}-CH(R)-\overset{\overset{\textstyle O}{\|}}{C}-O-Cell$	[14,15]

Alkyl isocyanate

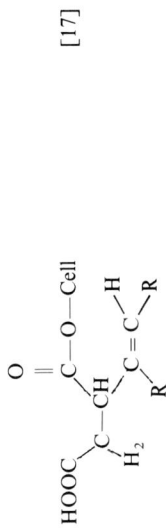

$R-N=C=O$

$C_{17}H_{35}$

$$R-N-C-O-Cell$$
(with H and O)

[16]

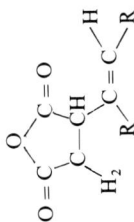

Alkenyl succinic anhydride

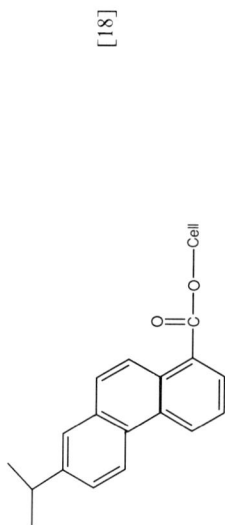

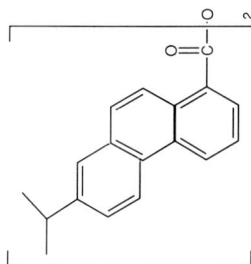

C_1-C_{16}

[17]

Rosin anhydride

[18]

migration during drying and must be sufficiently chemically reactive at these temperatures to undergo reaction with cellulosic hydroxyl groups.

The most commercially successful of these sizes are the alkyl ketene dimers (AKD) and the alkenyl succinic anhydrides (ASA). These are at opposite ends of the spectrum of reactivity and hydrolytic stability. AKDs are rather unreactive towards cellulose, except at elevated temperatures, but have good hydrolytic stability. The reverse is true for ASAs.

9.2   Reasons for changing to neutral sizing

The driving forces for papermakers to change from acid to neutral paper making are not always easy to establish, but probably the most important one is that neutral pHs allow the use of calcium carbonate, a relatively inexpensive and good quality filler (see Chapter 11). The availability of good quality carbonate in Europe is probably the reason why neutral paper making has developed more rapidly there than in North America. However, the situation in North America is changing quickly as a result of the increasing use and availability of precipitated calcium carbonate (PCC). PCC is produced by first calcining natural limestone to calcium oxide which is then converted to calcium hydroxide by the addition of water. Carbon dioxide, from the flue gas of the steam raising plant is then used to precipitate the calcium carbonate. The carbonate can, by correct control of the precipitation, be obtained in various crystalline forms, particle sizes and surface areas which give it significant advantages over natural calcium carbonate (NCC). Although inherently a more expensive product than (NCC), PCCs can be cost effective because they are produced on-site in satellite plants which do not require the slurry to be dried, and because they offer improved performance. They cannot, for obvious reasons, be used in the acidic rosin–alum environment.

A further consequence of the increased use of calcium carbonate is that there will be a trend towards higher levels of carbonate in recycled paper. Those waste based mills which are still operating under acidic conditions will therefore be increasingly forced to convert to neutral/alkaline systems.

Other important advantages of neutral paper making which are often quoted are improved permanence, reduced machinery corrosion and increased paper strength resulting from the increased swelling obtained at higher pH. However, these should more correctly be considered to be advantages which arise from operating at higher pH rather than from the change of sizing system.

Sizes which are now used in neutral and alkaline systems were therefore developed in response to the papermakers' need to move to neutral papermaking conditions. However, they represent an entirely new approach to the problem of introducing hydrophobic groups to fibre surfaces.

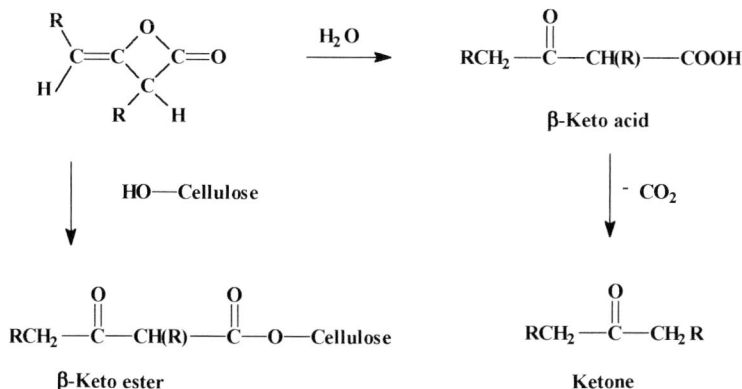

Figure 9.3 Reaction of alkyl ketene dimers ($R = C_{14}H_{29}$ to $C_{18}H_{37}$) with cellulose and water.

9.3 Alkyl ketene dimers

9.3.1 Synthetic developments

Alkyl ketene dimer sizing agents were a direct development of basic research carried out in the 1940s [19] which demonstrated that the parent molecule, diketene, could derivatise hydroxyl groups, in particular those of cellulose. Their selection as potential sizing agents for cellulose seems to be based on the work of Staudinger and Eicher in 1952 [20] who used diketene to acetoacetylate cotton in glacial acetic acid using a sodium acetate catalyst. Kirillova and Padchenko [21] later extended this work to a range of conditions and catalysts. However, these non-aqueous conditions are very different from those used in the papermaking process, and the reaction cannot be assumed to occur so readily in an aqueous system. Reactivity is discussed more fully in section 9.3.3. In 1947, Sauer [22] synthesised the first alkyl ketene dimer thus paving the way for Downey [14] in 1953 to develop the first alkyl ketene dimer containing long chain hydrocarbon alkyl groups for use in paper sizing. This work appeared in the patent literature in 1953 [14] and later (1956) in the open literature [23].

They are believed to form a direct covalent linkage with cellulose via β-keto ester formation. The general mechanism for this and the competitive hydrolysis reaction is shown in Figure 9.3.

9.3.2 Emulsion preparation and retention

Commercial AKDs are prepared from commercial stearic acid via the acid chloride:

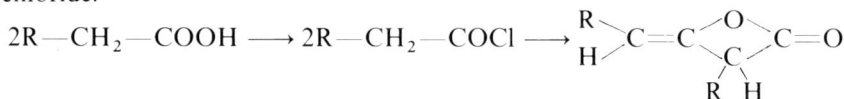

They are waxy, water-insoluble solids with a melting point around 50°C and are prepared as a stabilised emulsion by dispersion in a cationic polymer (normally cationic starch). They may vary from slightly to highly cationic with a dry solids content usually in the range 6% to about 15%. Typically around 30% of these solids would be made up of AKD and the remainder of cationic polymer. Small amounts of retention aids and surfactants may also be present. Particle size distributions are in the range 1–5 microns, and addition levels would be around 1–2% of the as-received emulsion based on fibre. This is equivalent to a level of 0.05–0.1% of pure AKD based on fibre.

The mechanism of retention is believed to be by heterocoagulation of the cationic size particles to the negatively charged fibre surface. This retention mechanism has been reviewed in detail elsewhere [24]. The charge character-istics of an AKD emulsion stabilised with a tertiary cationic starch as a function of pH are shown in Figure 9.4. There is a decrease in electrophoretic mobility as pH is increased, which is consistent with deprotonisation of the charged tertiary amino cationic group at high pH. The charge characteristics of bleached Kraft pulp are also shown in Figure 9.4, and this suggests that optimum retention would be found at around neutral pH. This has been found to be the case [25, 26]. The nature of the charged cationic group is clearly very important. A quaternary cationic group, for example, would not be expected to display the same decrease in mobility at high pH.

Sizing efficiency has also been shown to be dependent upon the chain length of the hydrocarbon groups. As the carbon number of the hydrocarbon chain increases from C-8 to C-14, sizing efficiency increases. It then levels off with only minor variations up to C-20 [27].

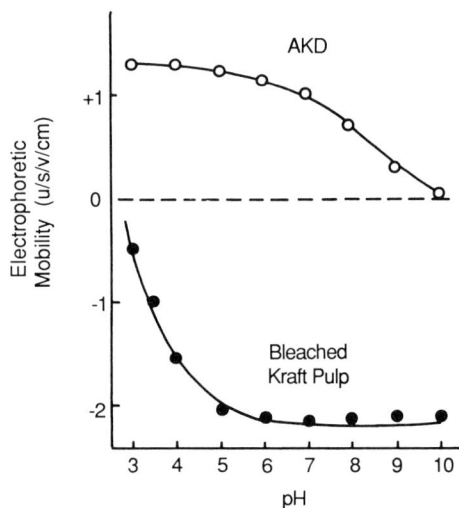

Figure 9.4 Effect of pH on electrophoretic mobility of AKD emulsion particles and a bleached Kraft pulp [30].

9.3.3 Reactivity

Diketene and alkyl ketene dimer derivatives have been shown to be surprisingly unreactive towards water and hydroxyl containing compounds even in homogeneous solution at relatively high temperatures [28]. The general reactivity of the higher analogues of diketene, the alkyl ketene dimers, would be expected to be lower than that of diketene because of the steric crowding of the reactive lactone ring by the two long chain alkyl groups. This has indeed been found to be the case for tetradecyl ketene dimer [26]. The rate of its hydrolysis and also its reaction with alcohols at neutral pH is very slow but, under basic conditions, it is much more rapid. The authors suggest that reaction between alkyl ketene dimers and cellulose in the papermaking system is likely to be slow, and in particular it is likely to be slower than the competitive hydrolysis with water.

Marton [29] studied the kinetics of cure and hydrolysis, and concluded that the rate of hydrolysis to form the ketone would be significant at the high temperatures experienced during drying. Marton also further demonstrated that the ketone hydrolysis product exhibited no sizing, thus confirming the work previously reported by Garner [25].

Evidence for direct β-keto ester formation with cellulose has been difficult to obtain. It was initially based on the observation that the dimer appeared not to be extracted by organic solvents and that, on treatment of the dimer-sized fibres with cuprammonium solution (a solvent for cellulose), an insoluble shell remained. This was presumed to be the reacted external surface of the fibre [23]. Research with ^{14}C-labelled dimers [30–32] has shown that, although some of the dimer is irreversibly bound to the fibres, varying amounts of the unreacted dimer can, in fact, always be extracted. The existence and extent of any reaction is therefore somewhat controversial, and attempts have been made to demonstrate β-keto ester formation by FTIR spectroscopy [33–35]. Pisa and Murkova [33] and Rohringer [34], using multiple internal reflectance infrared spectra of sheets of paper sized with alkyl ketene dimers, were unable to detect the presence of β-keto ester linkages but were able to detect the characteristic infrared bands associated with the unreacted dimer. Lindström [35], however, claims evidence for a β-keto ester linkage using similar methods. More recently, Bottorff [36] has used a ^{13}C labelled dimer and solid state ^{13}C NMR spectroscopy to demonstrate the formation of AKD-cellulose β-keto esters. He also showed that the AKD which was associated with calcium carbonate filler was transformed to the ketone hydrolysis product over time and that the rate of this transformation depended upon the type of calcium carbonate. Nahm [37] has also demonstrated that a reaction can be made to occur between cellulose and AKD under non-aqueous conditions using dimethylformamide as a solvent. However, it is difficult to extrapolate these results to the conditions of normal papermaking. The weight of evidence therefore suggests that β-keto ester formation is probably the important reaction involved in inducing sizing.

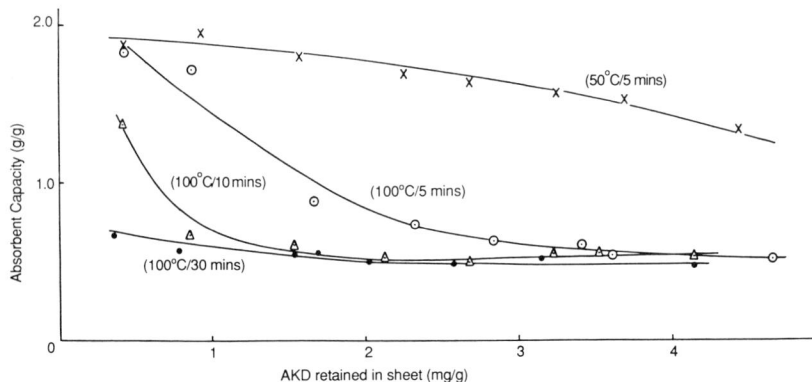

Figure 9.5 Effect of heat treatment on AKD sized handsheets (dried and conditioned at 23°C and 50% relative humidity before heating [30].

Isotopic (^{14}C) labelling techniques have proved to be very useful in investigating the reactivity of the dimers towards cellulose and water in paper prepared under standard handsheet conditions [30–32]. These techniques have demonstrated that heat treatment is essential for reaction to take place and for the development of sizing by alkyl ketene dimers [30, 31]. Figure 9.5 demonstrates the importance of heat treatment. Furthermore, the more severe the heat treatment the lower is the level of retained dimer which is necessary to induce good sizing. This is probably a distribution effect, and support for this has been obtained by autoradiography of sheets sized with ^{14}C-labelled AKD. This shows that diffusion and migration occur as a result of heating [26].

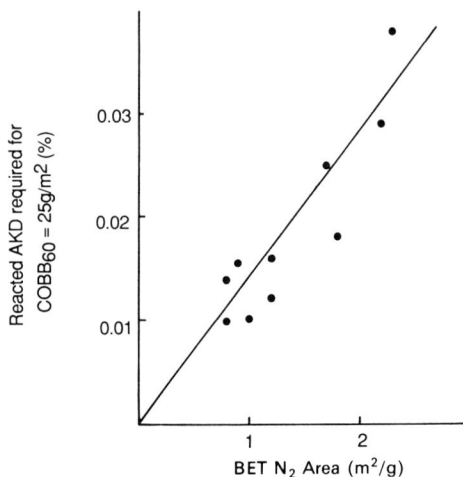

Figure 9.6 Correlation between reacted AKD and BET (N$_2$) surface area for different pulps [31].

The degree to which the dimer has become irreversibly 'bound' or reacted with the cellulose has been studied by solvent extraction methods [30, 31]. Sheets which had been sized with labelled compounds and then dried by different methods were extracted with a solvent capable of removing the unreacted AKD. Lindström [31] was able to show that the small amount of AKD required to give a $COBB_{60}$ of $25\,g\,m^{-2}$ after removal of unreacted dimer by solvent extraction (0.008–0.038%), showed a strong correlation with the BET-surface areas of the pulps used (Figure 9.6). Using these results together with surface balance measurements, a surface coverage of 4% of planar oriented monolayer was calculated to be necessary to give this level of sizing.

Because of the rather slow rate of reaction, catalysis has also been studied. HCO_3^- ions and polyamideamine–epichlorohydrin resin (PAMAM) have been shown to exert a strong catalytic effect [38]. A trimolecular mechanism for the catalytic effect of HCO_3^- ions has been proposed [38].

One of the most important features of AKD sizing is that the development of sizing takes place after drying and during reel-up. This has been studied recently [39] and it has been shown that sizing development may take place over a period as long as 10 days and that the ultimate sizing level is a function of the drier temperature (Figure 9.7). Higher storage temperatures were also found to have a large effect upon the rate of sizing development. Greater levels of reaction were observed to occur during storage.

Second generation AKD sizes now exist [40] in which specifically designed cationic polymers have been incorporated into the AKD emulsion preparation as promoters to improve the retention of the dimer and also to improve the cellulose reaction kinetics. However, little detail is available in the basic research literature on the mechanism by which such additives

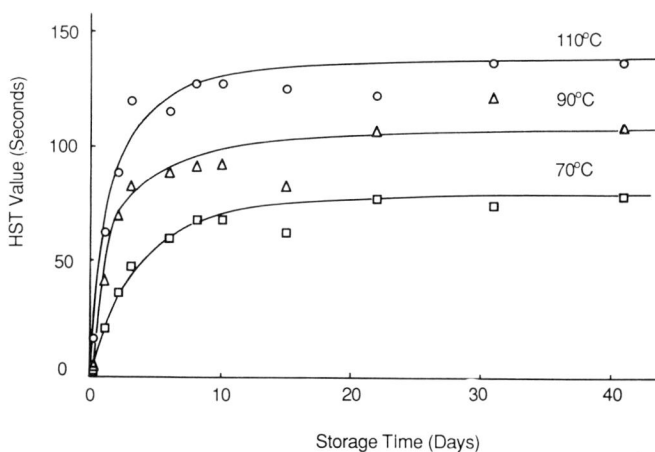

Figure 9.7 Effect of drying temperature and subsequent storage time on AKD (0.1%) sized paper [39].

impart improvements. Some practical studies have also been carried out in which the chemical environment has been varied and promoter-free and promoter-containing AKDs have been used. The most effective combination was found to be promoter-free AKDs with modified polyethylene imine and an anionic polyacrylamide as the retention aid system [41]. However, this finding was made on a commercial paper machine system and may not be universally applicable.

9.3.4 *Practical considerations*

A number of process variables influence AKD sizing, the most important of which are probably pH and drying conditions. It is common to dry sheets to very low moisture contents before the size press (typically 3–5% but sometimes as low as 1–3%). This assists greatly but, even under these conditions, there is still considerable off-machine development of sizing [27]. AKD sizes are generally used over a pH range of 6–9, but are more effective at the higher end of this range [27]. The effect of alum is more complex, for there is evidence that it has an adverse effect which is only in part due to the effect of its low pH [26]. Pulp type is also important. In general, unbleached pulps are easier to size than bleached pulps. High α-cellulose pulps are extremely difficult to size and require as much as ten times the amount of reacted AKD to produce similar levels of sizing to those of Kraft pulps [27]. This effect cannot be explained in terms of surface area.

The presence of fillers usually increases the amount of AKD required. As reaction with filler surfaces is not possible, sizing is probably brought about by the unreacted AKD or the ketone hydrolysis product. A proportionally greater amount of size may therefore be needed.

Some AKD sizing is carried out by application at the size press. This is still relatively unusual but is growing in popularity. Sizing by this method is more likely to be restricted to the surface and it is often used in addition to wet-end sizing rather than in lieu of it.

Finally, there have been a number of studies into the effect of AKD sizing on the surface properties of paper. There is some evidence that neutral sizes tend to give rise to a more slippery sheet than rosin–alum sizing. This can have important implications for those converting operations where lower frictional resistance is undesirable. Marton [29] has demonstrated that the presence of dialkyl ketones derived from the hydrolysis of AKD on the surface of paper decreases the friction observed between contacting paper surfaces. There is also some evidence that AKD may assist in suppressing the capillary movement of aqueous adhesives when corrugated starch adhesive is used in board by reducing the extent of dehydration of the adhesive [42].

Borch and Miller [43] used electron spectroscopy (ESCA) to study size distribution on the surface of printing papers. They showed that there was an increase in the hydrophobicity of the surface of sized sheets and that this corresponded to a decrease in the oxygen to carbon ratio. The authors used this as a measure of sizing and were able to demonstrate that sizing was well distributed at the resolution level of the electrophotographic printing techniques.

There is also evidence that the decrease in surface energy of the paper surface which is brought about by AKD sizing has a detrimental effect upon polyethylene adhesion strength [44].

9.4   Alkenyl succinic anhydride

9.4.1   *Synthetic development*

The development of alkenyl succinic anhydrides as sizing agents took place in 1974 [45]. Like AKDs, they are able to undergo reaction with cellulose and water and the reaction paths are shown in Figure 9.8. The ASAs are quite reactive molecules and can promote sizing without heat treatment [46]. They are prepared from 1-alkenes (alpha olefines) by catalytic isomerisation [47] followed by an addition reaction with maleic anhydride (the ene reaction). The reaction scheme is shown in Figure 9.9. A full discussion of the stereochemistry and mechanism of this reaction is available elsewhere [48, 49], but it is worth noting that for each positional isomer there are eight stereoisomers possible because of the two asymmetric centres and the

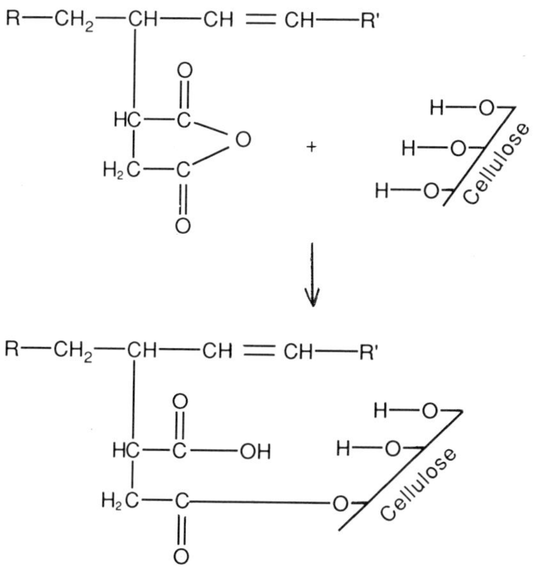

Figure 9.8   Reaction of alkenyl succinic anhydride with cellulose.

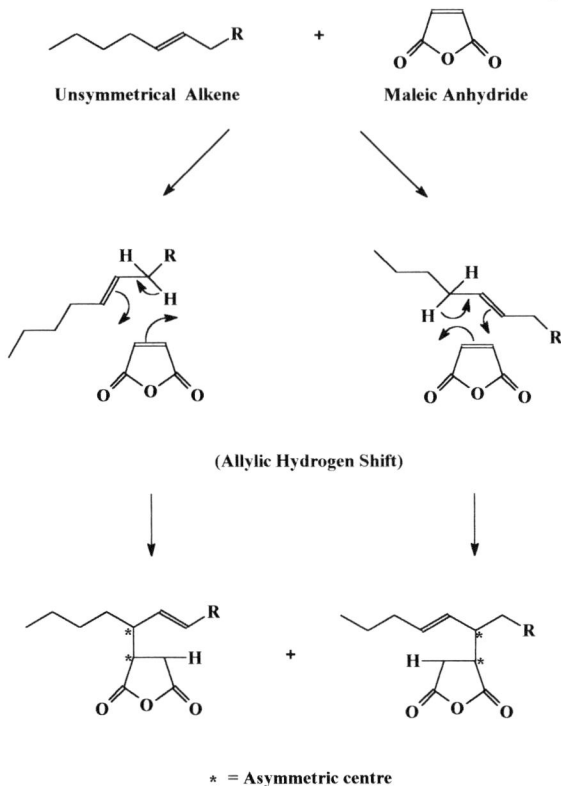

Figure 9.9 Mechanism of the ene reaction for the formation of alkenyl succinic anhydride.

presence of a double bond. As a consequence of the concerted mechanism shown in Figure 9.9 an unsymmetrical alkene can give two positional groups of ASA isomers, and sixteen isomers are therefore possible for ASAs synthesised from a single unsymmetrical alkene. This taken with the fact that the double bond in commercial alkenes has been moved along the chain by catalytic isomerisation, and also that the alkenes are of mixed chain length means that upwards of 300 isomers are likely to be present in a commercial ASA.

Unlike the AKDs, which are derived from fatty acids, the ASAs are petrochemical based, and this has implications for their economics, particularly during times of high oil prices. The location of the double bond in the alkene is important and it has been shown that the internal alkenes are much more effective than α-olefins [17]. The probable explanation for this is that the ASAs derived from α-olefins, being solids at room temperature, require higher temperatures for emulsification than the isomerised ASAs derived from isomerised olefins. This is likely to accelerate rapidly the unwanted hydrolysis of the size. However, this is speculative and a definitive explanation is awaited.

9.4.2   Emulsification

Anhydrides made from alkenes with a carbon number of C-16–C-18 are usually used commercially. Alkenes with a carbon number lower than C-14 tend to give anhydrides which are less effective [17]. ASAs prepared from alkenes with a carbon number of more than C-20 are solids at room temperature and are less suitable for emulsification. Emulsions are usually prepared using cationic starch as the stabilising polymer. Starch concentrations in the emulsions are usually in the range 1–2%, and the ratio of starch to ASA is around 3:1. Emulsification is carried out on-site using either a low-shear venturi system or a high-shear turbine system. In the former, a surfactant (usually around 5% based on ASA) is often used, although some consider the surfactant to be detrimental to sizing. In the high shear turbine system less surfactant (typically around 1% based on ASA) is used. Regular maintenance of the system is required (typically monthly) and it is usual to have a back-up instrument. Particle sizes in the range 1–2 μm are aimed for and can usually be easily achieved.

9.4.3   Hydrolytic stability

The hydrolytic stability of ASA emulsions has been studied by various workers [50, 51] and some typical results are shown in Figure 9.10(a, b). At an initial pH of around 7, and without pH adjustment, the hydrolysis is complete in around 30 hours, the final pH of the emulsion being around 3.5 due to the formation of the weakly acidic dicarboxylic acid. The hydrolysis displays a short induction period followed by an accelerating rate. The acceleration is difficult to explain on the basis of particle size considerations, as the hydrolysis product is reasonably soluble in water at these concentrations and would be expected to be dissolved from the particle–water interface during hydrolysis. This would lead to a gradual decrease in the total surface area of the emulsion and a decrease in the rate of hydrolysis. It is also not possible to explain the acceleration in terms of an autocatalytic effect, for although anhydride hydrolysis can be both acid and base catalysed, their catalysis is insignificant even at a pH of 3.5 [52, 53]. The most likely explanation seems to be in terms of the protective action of the starch molecule at the surface of the emulsified ASA particle which may act as a barrier to the transport of water to the particle surface. As water diffuses through this barrier, hydrolysis occurs, the emulsion becomes destabilised and the rate increases. An alternative explanation is that the hydrolysis product dissolves in the ASA particle and, in doing so assists in the transport of water into the particle interior, hence increasing the rate.

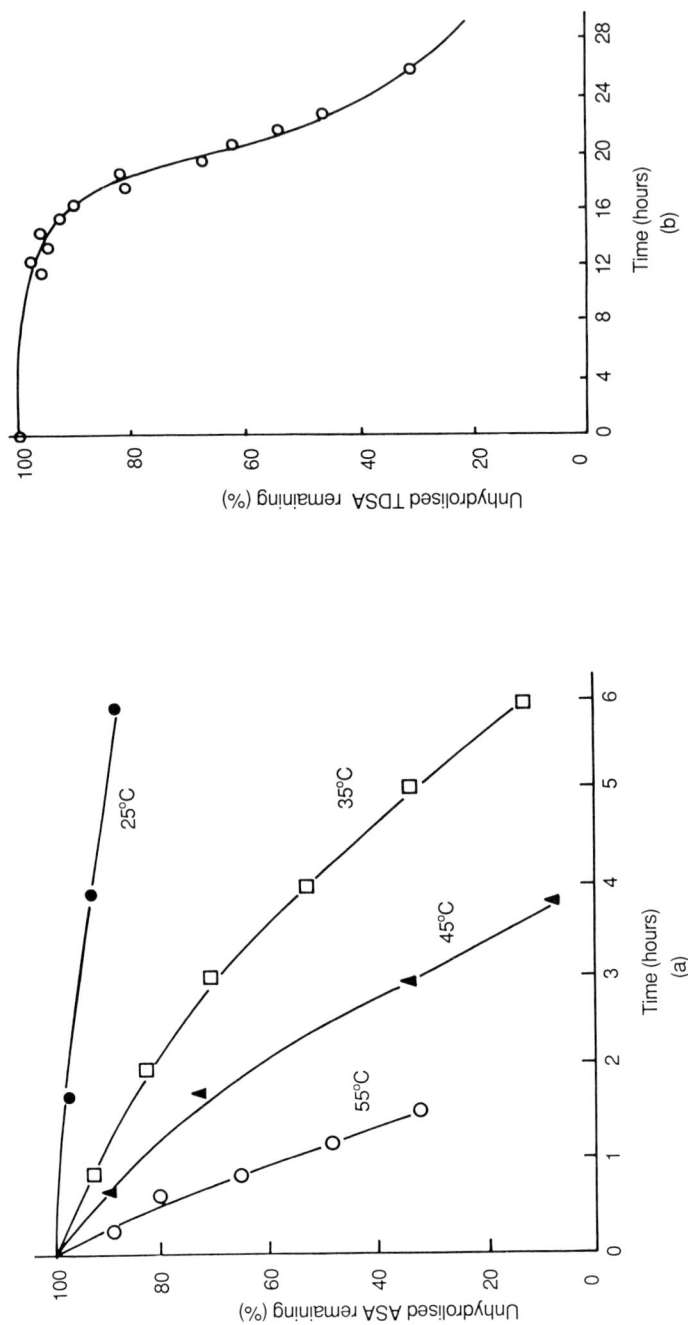

Figure 9.10   Rate of hydrolysis of (a) commercial ASA at initial pH of 3.5 [43], and (b) tetradecenyl succinic anhydride at initial pH of 7.2 [44].

Work with ^{14}C-labelled ASA has shown that sizing can be induced merely by conditioning the sheets for 24 hours at 20°C and 50% RH, no heat treatment being necessary [46]. This contrasts sharply with AKD sizing, in which sizing is only developed after the sheet has been subjected to heat treatment [26, 30, 31].

McCarthy and Stratton [54, 55] have demonstrated that the dicarboxylic acid is inhibitory to sizing. Roberts and Wan Daud [46] also showed that solvent extraction of ASA-sized sheets usually led to an increase in sizing, and explained this in terms of the inhibitory effect of hydrolysed excess ASA. Confirmation of the de-sizing nature of the acid was obtained by Roberts and Wan Daud [46] who performed an experiment in which sheets were sized with ASA at pH 7 and the unreacted and/or hydrolysed ASA extracted with chloroform. The amount of reacted ASA was 0.31 mg/g and the sheets had an HST value of 207 seconds. Dicarboxylic acid was then applied at different levels of addition by a solvent application procedure from toluene. A progressive decrease in sizing was observed (Figure 9.11), and the sizing could be completely restored to its original level by solvent extraction of the applied acid. It seems probable that the dicarboxylic acid produced by hydrolysis is able to become adsorbed to previously reacted alkyl groups in a way which allows the orientation of its polar carboxyl groups away from the fibre surface and towards advancing water molecules. This behaviour contrasts with that of AKD sizing, where no increase in sizing is observed after solvent extraction. In the case of AKD the rate of hydrolysis is much slower, but there is also no

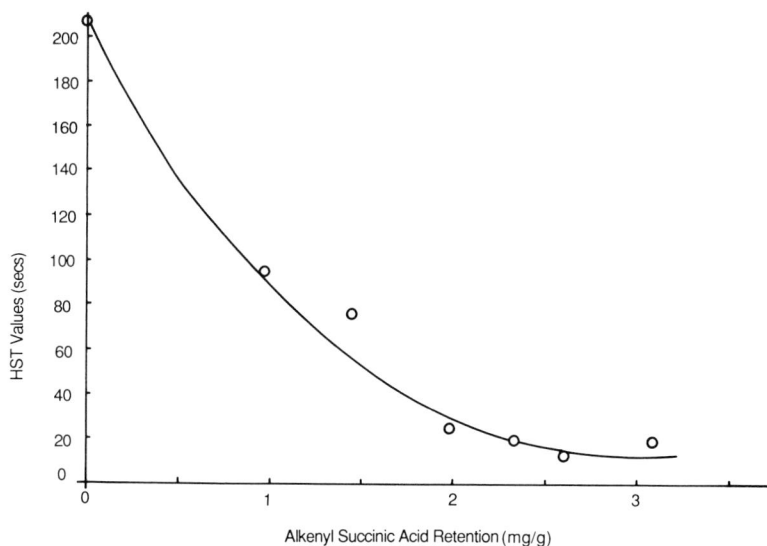

Figure 9.11 Effect of alkenyl succinic acid addition on sizing of ASA sized handsheets [46]. (Note initial level of reacted ASA = 0.31 mg/g).

evidence that the ketone produced by hydrolysis acts as a desizing agent [25, 29].

The effect of pH on the degree of reaction and upon the sizing has been studied [56]. Ink penetration times were shown to be highest at pH 7 and lowest at pH 11 for equivalent levels of retained ASA. The sizing at pH 4 was slightly lower than that at pH 7. These results were explained in terms of the competitive hydrolysis, which is considerably faster at pH 11 than at pH 7 and probably marginally faster at pH 4 than at pH 7.

It is likely that residual unreacted ASA in a sheet will slowly hydrolyse to its dicarboxylic acid and may then contribute a de-sizing effect. Optimum sizing in the ASA system is therefore likely to be achieved by maximising the level of reaction with cellulose and minimising the degree of hydrolysis. It is also likely that increasing the amount of ASA used beyond a certain threshold level will not be beneficial.

9.4.4 *Practical consideration*

There are a number of practical considerations when using ASA as a size. Hydrolysis of the size, as has been discussed earlier, must be avoided. Emulsion storage time should therefore be minimised or eliminated. The ASA should also be added at a point which keeps contact time with the stock to a minimum. It is also important to retain as much of the size and also the fines, onto which much of the size is adsorbed, as possible in order to prevent recirculation of hydrolysed size around the whitewater system. Hydrolysed ASA in the whitewater system tends to give pitch and deposit problems, probably as a result of formation of insoluble calcium salts of the dicarboxylic acid. The tackiness of these soaps has been related to a number of variables such as pH, retained metal and the amount of free acid.

9.5   Comparisons between AKD and ASA sizing

The main difference between the two sizes is the lower reactivity of the AKDs. Whilst undersirable, this does have advantages in terms of the stability of the aqueous emulsion. AKD emulsions are stable towards hydrolysis for several months [25, 27], which means that they can be delivered by tanker and stored for a reasonable period before use. ASA emulsions, on the other hand, have a very poor hydrolytic stability (see section 9.4.3) and must be prepared on-site immediately prior to use (section 9.4.2). In addition, the products of hydrolysis differ in their effects. The product of hydrolysis of AKD is a ketone which, although having a small positive sizing effect in its own right, is not as effective weight for weight as the covalently bound AKD [25]. Excess unreacted AKD in the sheet, whilst undesirable, is not a particular problem as far as sizing is concerned. The hydrolysis product of ASA, on the other hand, has a seriously detrimental effect on sizing if it is retained in the sheet. This effectively limits the extent to which sizing can be achieved using ASA. Excess unreacted ASA will be hydrolysed by equilibrium moisture in the sheet, and attempts to increase sizing by increasing the level of addition of ASA may therefore be self-defeating.

Levels of reaction which can be achieved with each of these sizing systems are therefore important, and this is a subject which has given rise to a certain amount of controversy in the literature [26, 28, 31, 34, 35]. There are those who maintain that any reaction at all between cellulose and AKD is unlikely under papermaking conditions [34], and there are those who maintain that very high levels of reaction can be obtained [31]. Close scrutiny of the experimental methods and conditions, where available, shows that AKD is in fact very unreactive towards cellulose unless high pH and extreme drying temperatures and times are used [25, 26, 30]. Most commercial papers made at pHs of around 7–8 will contain a large proportion of unreacted AKD which will presumably eventually be converted into its ketone hydrolysis product. However, the treated AKD gives extremely high levels of sizing. The biggest technological challenge at the moment in research into AKD sizing is to find a method of application of these dimers which allows high levels of reaction at lower levels of addition. Levels of reaction in ASA sizing are equally important but for another reason, namely that the unreacted material can hydrolyse quickly and desize the sheet.

One of the most important differences between AKD and ASA sizing is that, in the case of the latter, reaction with cellulose and sizing can be obtained without the sheet temperature having to be raised above ambient. This has important implications for the rate of development of sizing. Sheets sized with ASA have developed their sizing immediately after leaving the drier section of the machine. AKD sized paper tends to develop sizing over a period of time (often weeks) after leaving the drier section. The final level of sizing for AKD sized paper is therefore often difficult to gauge, and this is obviously a commercial disadvantage.

The fundamental difficulty with reactive sizes is that there will always be competition between hydrolysis and reaction with cellulose. Those sizes which are very reactive towards cellulose are also susceptible to hydrolysis and those which are not susceptible to rapid hydrolysis are rather unreactive towards cellulose. This problem can never be completely overcome whilst paper is made in an aqueous environment. However, some improvements in the chemistry can be visualised. A size which was more reactive than AKD, but whose hydrolysis product did not desize would have obvious advantages.

What has emerged from the recent series of publications on reactive sizes is that they are clearly effective, they are relatively simple to use, and their underlying chemistry is much more straightforward than that of rosin and alum. There has been a major, and probably irreversible, shift towards them over the past fifteen years and, as the move to neutral and alkaline paper making shows no signs of stopping, they are here to stay.

9.6 Neutral rosin sizing systems

In the past it has been difficult to use rosin as a sizing agent under anything other than acid conditions. This is due to the acidity of aluminium sulphate solutions. However, more recently, neutral pH rosin sizing has become possible by either manipulating the rosin–alum system (see chapter 8) or by substitution of polyaluminium chloride (PAC) for aluminium sulphate. Some general articles on the use of polyaluminium chlorides with rosin for neutral sizing have been published [57–59] but little information is available on the mechanism of sizing.

9.6.1 *Polyaluminium chloride*

Polyaluminium chloride is the name given to a series of polymeric aluminium chlorides in which hydroxyl ions are substituted for some of the chloride ions. It is also known as aluminium chlorohydroxide, aluminium hydroxychloride and polyaluminium hydroxychloride. They can be prepared in a number of ways. Aluminium chloride can be hydrolysed with a base such as NaOH, Na_2CO_3, NH_4OH or lime. Alternatively HCl can be reacted with aluminium hydroxide at high pressure. They have the general formula $Al_n(OH)_mCl_{3n-m}$, and their degree of polymerisation varies and is a function of the hydroxyl concentration. In many ways the chemistry of solutions of aluminium chloride is similar to that of aluminium sulphate. Both form the octahedral hexahydrated complex aluminium ion. $\{Al(H_2O)_6\}^{3+}$, in solution in which six water molecules are co-ordinated to the aluminium ion. This complex is highly sensitive to pH and will form polymeric ions of the general formula given above. In the case of aluminium chloride these polyaluminium species remain soluble and retain their cationic charge to a higher pH range than aluminium sulphate. The simplest polymeric species is the dimer, $\{Al_2(OH)_2(H_2O)_8\}^{4+}$. The PACs are characterised by their aluminium to chlorine ratio or, alternatively, by their basicity. The latter is defined as follows:

$$\% \text{ Basicity} = [OH^-]*100/3[Al^{3+}]$$

Where $[OH^-]$ and $[Al^{3+}]$ are the total molar concentrations of hydroxide and aluminium ions.

As basicity increases, the degree of polymerisation and the cationic charge of the polynuclear species increases. However, the charge density, as expressed by the number of charges per aluminium atom, decreases. PACs with a basicity of between 30 and 50% have been found to be effective and to give the ideal distribution of polynuclear aluminium species [60]. These highly charged polynuclear aluminium species are effective as retention and drainage aids, and may also be used to carry out rosin sizing at neutral pH.

9.6.2 Neutral rosin sizing with PAC

Polyaluminium chlorides are generally used with dispersed rosin sizes which do not dissociate significantly at neutral pH. There is only very limited information available in the literature on the application of this sizing system to commercial paper making systems. However, one approach which has been found to be successful is the simultaneous addition of dispersed free rosin and PAC to the stuff box, cationic starch to the stuff box recirculation line, precipitated calcium carbonate to the fan pump and the use of an anionic retention aid [60]. Levels of addition have been reported to be around 0.2–0.3% rosin (based on fibre) and 1–1.2% of the as-received polyaluminium chloride product and sizing values of between 30 and 80 seconds HST have been obtained.

References

1. Olsson, I. and Pihl., *Svensk Pap.*, **55** (1952), 233.
2. Hoyland, R.W., *Proc. Fund. Res. Symp. Oxford* (1977), 557.
3. Fox, H.W., Zisman, W.A., *J. Colloid Sci.*, **7**(4) (1952), 428.
4. Hoernschmeyer, D., *J. Phys. Chem.*, **70** (8) (1966), 2628.
5. Bernett, M.K., Zisman, W.A., *J. Phys. Chem.*, **63** (8) (1959), 1241.
6. Tappi Monograph, *Sizing of Paper*, ed. W.F. Reynolds (1989), 133–154.
7. Strazdins, E., *Tappi*, **64** (1) (1981), 31.
8. German Patents: 2,423,651 (1974); 2,611,827 (1976); 2,611,746 (1976).
9. US Patent 4,123,319 (1978).
10. British Patent 954,526 (1964).
11. Canadian Patent 770,079 (1967).
12. US Patents 3,102,064 (1963); 3,409,500 (1968); 3,455,330 (1969); 4,207,142 (1980).
13. Serota, S. *et al.*, *Tappi*, **63** (9) (1980), 92.
14. Downey, W.F. to Hercules; US Patent No. 2,627,477 (1953).
15. US Patents 2,785,067 (1957); 2,762,270 (1956); 2,856,310 (1958); 2,865,743 (1958); 2,961,366 (1960); 2,986,488 (1961) 3,483,077 (1969).
16. US Patents 3,050,437 (1962); 3,589,978 (1962); 3,492,081 (1970); 3,310,460 (1967); 3,627,631 (1971); 3,499,824 (1970); 3,575,796 (1971).
17. US Patent 3,821,069 (1974).
18. Dumas, D.H., *Tappi*, **64** (1) (1981), 43.
19. Boese, A.B., *Ind. Eng. Chem.*, **32** (1940), 16–22.
20. Staudinger, H. and Eicher, T., *Makromol. Chem.*, **8–10** (1952–53), 261–279.
21. Kirillova, G.N. and Padchenko, G.O., *Zhurnal Prikladnoi Khimii*, **37**, (1964), 918.
22. Sauer, J.C., *J. Amer. Chem. Soc.*, **69** (1947), 2444–2448.
23. Davis, J.W., Robertson, W.H. and Weisgerber, C.A., *Tappi*, **39** (1956), 21–23.
24. Lindström, T., *Trans. 9th Fundamental Research Symposium*, Cambridge, England, Baker, C.F. and Punton, V.W., eds. Mech. Eng. Publ. Ltd., London (1989), 311–435.
25. Garner, D.N. Ph.D. thesis, University of Manchester, England (1984).
26. Roberts, J.C., Garner, D.N. and Akpabio, U.D., *Proc. 8th. Fund. Res. Symp.*, Oxford (1985), vol. 2, Mech. Eng. Publ., London (1985), 815–837.

27. Tappi Monograph, *Sizing of Paper*, ed. W.F. Reynolds, Ch. 2, (1989), 33–50.
28. Roberts, J.C. and Garner, D.N., *Cell. Chem. and Tech.*, **18**, (1984), 275–282.
29. Marton, J., *Tappi*, **73** (22) (1990), 139–143.
30. Roberts, J.C. and Garner, D.N., *Tappi*, **68** (4) (1985), 118–121.
31. Lindström, T., Soderberg, G. and Obrien, H., *Nordic Pulp and Paper J.*, **1**, 26–42 and ibid. **2**, 31–45 (1986).
32. Lindström, T., *Proc. XXI Eucepa Conf.*, Torremolinos, Spain (1984).
33. Pisa, L. and Murkova, E., *Papir Celluloza*, **36** (2) (1981), v15–v18.
34. Rohringer, P. *et al.*, *Tappi*, **68** (1) (1985), 83–86.
35. Lindström, T., Discussion contribution to reference 26.
36. Bottorff, K.J., *Tappi*, **77** (4) (1994), 105–116.
37. Nahm, S.H., *J. Wood Chem. Tech.*, **6** (1) (1986), 89.
38. Lindström, T. and Soderberg, G., *Nordic Pulp and Paper Res. J.*, **2** (1986), 39–45.
39. Roberts, J.C. and Yajun, Z., *Proc. PIRA Symp. on Neutral Papermaking*, Stratford, England (1990).
40. Watson, L.F., *PIMA*, **70** (9) (1988), 36–38.
41. Guth, K., Lorz, R. and Scholz, R., *Wochenbl. Papierfabr.*, **115** (22) (1987), 1014–1018.
42. Inoue, M. and Lepoutre, P., *Nord, Pulp. Pap. Res. J.*, **4** (3) (1989), 206–209.
43. Borch, J. and Miller, A.G., *Proc. 10th Cellulose Conf. 1988*, Syracuse, New York, 1429–1441.
44. Borch, J., *Tappi*, **65** (2) (1982), 72–73.
45. Wurzburg, O.B., US Patent No. 3,821,069 (1974).
46. Roberts, J.C. and Wan Daud, W.R., *Proc. 10th Cellulose Conference*, Syracuse, June 1988.
47. US Patent No. 3,641,184 (1972).
48. Benn, F.R., Dwyer, J. and Chappell, I., *J. Chem. Soc.* (Perkin II) (1977), 533–535.
49. Sublett, B.J. and Bowman. N.S., *J. Org. Chem.*, **26** (1961), 2594–2595.
50. Wasser, R.B., *Proc. Tappi Papermaker's Conf., Denver*, (1985), 17–20.
51. Wan Rosli Wan Daud, Ph.D. thesis, University of Manchester (1988).
52. Eberson, L., *Acta Chem. Scand.*, **18** (2) (1964), 534–542.
53. Bunton, C.A. *et al.*, *J. Chem. Soc.*. (1963), 2918–2926.
54. McCarthy, W.R., Ph.D. thesis, Institute of Paper Chemistry, Appleton, USA (1987).
55. McCarthy, W.R. and Stratton, R.A., *Tappi*, **70** (12) (1987), 117.
56. Roberts, J.C. and Wan Daud, W.R., 'Wood Processing and Utilisation', *Proc. Cellucon Japan Conf.*, eds. Kennedy, J.F., Phillips, G.O. and Williams, P., Ellis Horwood, Chichester, England, (1989), 133–142.
57. Faber, W. and Simonetti, A., *Wochenbl. Papierfabr.*, **118** (7) (1990), 255–256.
58. Wortley, B., **72** (1) (1990), 45–47.
59. Gerischer, G., *Wochenbl. Papierfabr.*, **115** (15) (1987), 655–658, 660–661.
60. Colusardo, A.R., *Tappi*, (10) (1990), 71–78.

10 Dyes and fluorescent whitening agents for paper

S.G. MURRAY

10.1 Introduction

Paper today still remains the most important information bearer despite the computer age and electronic mail. Striking presentations invariably use colour as the most effective means of attracting our attention. Paper dates back to ancient times (e.g. Egyptian papyrus) and its colouring began as long ago as the Middle Ages when inorganic pigments such as Prussian Blue and Ultramarine Blue were used. When synthetic dyes made an appearance after the discovery of Perkin's Mauve in 1856 [1], the versatility of dyes rapidly became realised.

Paper derived from wood became of importance only relatively recently; previously, paper was made from rags. Following inorganic pigments, the colouring of paper was initially with textile dyes. Direct dyes normally applied to cotton were tried because paper, like cotton, is cellulosic. Direct dyes applied to cotton require boiling conditions and additions of salt, whereas paper dyeing is generally a cold process and dyes applied to paper pulp have more access to the cellulose fibres because of their more open structure. Many basic dyes were found to be suitable (e.g. Bismark Brown, Malachite Green, Rhodamine, Auramine, etc.). They are economical and have very bright colours, but have poor lightfastness. Some acid dyes were also used, but these have poor fastness and give coloured backwaters, thus wasting a lot of dye, unless fixatives are used.

As paper production greatly increased, so the market for speciality dyes and fluorescent whitening agents (FWAs) was realised, and in 1981 worldwide consumption of dyes in paper was estimated by Martin [2] at 37 000 tonnes (in powder form).

Direct dyes now account for about 60% of the paper market, basic dyes $c.$ 30%, then acid dyes and pigments make up most of the remainder. Dyes are applied at as little as 0.005% (w/w) or less for very pale shades and tints, or as much as 10% (solid)–20% (liquid) for dark shades and blacks.

This chapter will attempt to explain why these compounds are coloured and why they are applied to paper.

10.2 Basic concepts of colour

It would be beyond the scope of this chapter to discuss colour technology in detail, and the reader is directed to literature elsewhere [3].

10.2.1 *Light absorption and reflection*

Light is electromagnetic radiation and can therefore be considered as waves of energy with a certain frequency and wavelength. The light we see is only a small portion of the whole electromagnetic spectrum with wavelengths in the range between 400 and 700 nm. When light is incident on a surface, some may be absorbed by dye or pigment molecules; the rest is reflected, and is perceived by our eyes as colour. If all the incident visible light is reflected completely, it appears white, whereas if all is absorbed it appears black. If a constant fraction of the light is reflected throughout the whole visible spectrum, then we see grey. Black, white and grey are *achromatic colours*. However, if some of the light is only partially absorbed, then *chromatic colours* are observed. The light is absorbed over a band of wavelengths and the position and shape of the absorption band determines the colour's characteristics (brightness, hue, etc.).

For example, if light is absorbed in the short wavelength region, 435–480 nm (blue light), then that reflected appears as yellow. The reflected light combines to give *additive mixtures*, such that the more that is reflected the lighter the colour. Table 10.1 shows the colours perceived for a variety of absorption bands. Green dyes are exceptions as they have two absorption bands, one in the red region and one in the blue, and this is often difficult to achieve. To obtain greens, blue and yellow mixtures are often applied. Such colour mixtures are *subtractive mixtures*, since less light is reflected to give darker colours.

The shape of an absorption band is as important as its position in the spectrum. The band will normally have a wavelength of maximum absorption (λ_{max}) and the smaller the band-width, the brighter and purer the colour.

Table 10.1 Colours of absorbed wavelength bands and colours perceived

Band of wavelengths absorbed (nm)	Colour of light absorbed	Colour of light reflected (perceived)
400–435	Violet	Greenish yellow
435–480	Blue	Yellow
480–490	Greenish blue	Orange
490–500	Bluish green	Red
500–560	Green	Reddish violet
560–580	Yellowish green	Violet
580–595	Yellow	Blue
595–605	Orange	Greenish blue
605–750	Red	Bluish green

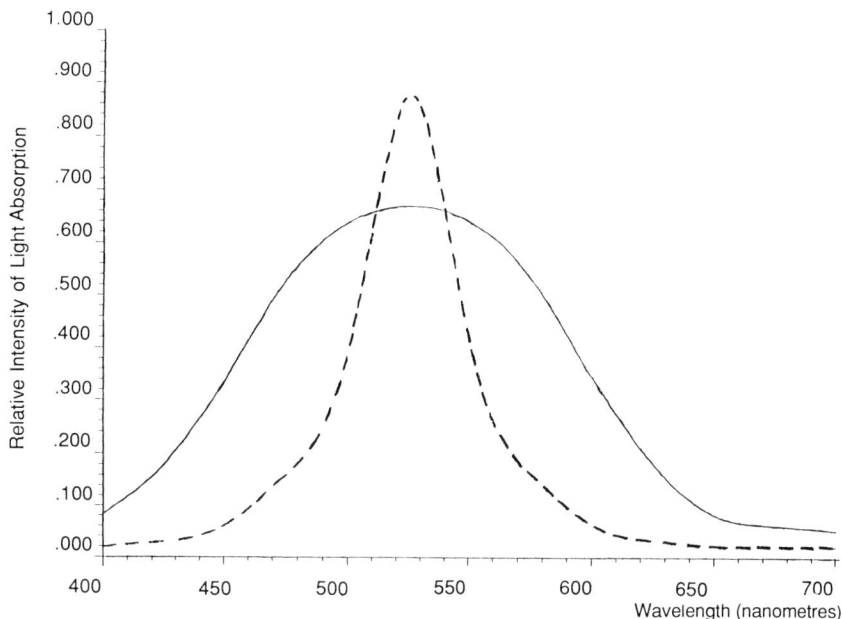

Figure 10.1 Absorption bands of two hypothetical reddish violet dyes. (— dull reddish violet; — bright reddish violet).

Thus in Figure 10.1, the two absorption bands have the same λ_{mas} (reddish violet, 525 nm) but the broad band gives a much duller shade, and the narrow band gives a shade which appears much brighter and more brilliant. Such effects are caused by the rigidity of the dye molecule and molecular vibrations (cf. section 10.2.2). Duller colours such as brown typically have broad absorption bands, sometimes with no discernible λ_{max}.

10.2.2   Colour and constitution

Reviews and books on this subject can be found in [4–9]. They deal with colour and constitution by various methods, the most popular being the molecular orbital (MO) method.

Generally, organic dye molecules contain atoms bound together by means of bonding electrons. A σ bond between two atoms consists of two shared electrons which lie in an orbit surrounding both participant atoms. However, some atoms may also be bonded together by a π bond, thus forming a double bond. The electrons in a π bond may delocalise to form conjugated systems (e.g. benzene). There is also the possibility of lone pairs of electrons being present on certain atoms (e.g. nitrogen atoms) which are non-bonding (n electrons) but which may also partially delocalise by interaction with π molecular orbitals to form specific molecular orbitals of different energy levels. In dye molecules the systems of delocalised π electrons may be very extensive over many atoms. The electrons may also exist in other specific orbitals which are termed *antibonding* when the molecule is momentarily in an *excited state* (of higher energy). The energy difference between the excited state and the unexcited or *ground state* is the criterion which determines the colour of the molecule. If this difference in energy, ΔE, corresponds to that of an incident photon of light, then the energy may be absorbed by the molecule, giving rise to an excited state.

The excited molecule may lose this extra energy in a number of ways. Normally, there is a possibility of the energy being dissipated by means of vibrations of the bonds between the atoms (observable by the formation of heat) until the original ground state is achieved. However, if the molecule has a rigid structure, such vibrational relaxations may be restricted and it is possible instead for such a molecule to emit a photon of energy as light, and fluorescence is observed. There are other ways for energy to be lost such as internal conversions, phosphorescence and degradation. If the molecule contains atoms with non-bonding lone pairs of electrons (e.g. nitrogen atoms), then these too may be excited. Excitation usually takes place by the most energetic electron occupying a molecular orbital in the ground state (either a non-bonding electron of a lone pair or a π electron) moving to the least energetic anti-bonding orbital.

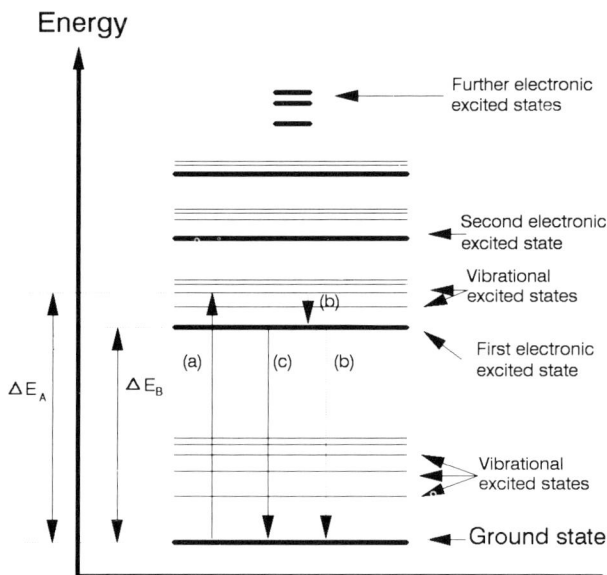

Figure 10.2 Light absorption (a), vibrational relaxations (b) and fluorescence (c) of organic dye molecules.

Figure 10.2 shows how excitation occurs from the lowest vibrational levels of the ground state to one of the vibrational levels of an electronic excited state by the absorption of light. Then energy is usually lost by vibrations of the bonds or through fluorescence. In fluorescence the wavelength of the radiated light is longer than that of the absorbed light (since $\Delta E_B < \Delta E_A$). Energy (E) and wavelength (λ) are related through the equation $\Delta E = hc/\lambda$ where h is Planck's constant and c is the velocity of light. Thus the smaller ΔE, the longer the wavelength of light required for absorption. To make ΔE smaller, it is possible to have electron donating groups present in the molecule (e.g. amino, alkylamino, hydroxy etc.) which can effectively make the most energetic occupied ground state molecular orbital of higher energy and thus the energy required for excitation less. In many dye molecules, this results in a much more intense absorption band at longer wavelengths. The effect can be further enhanced by the inclusion of further electron donating groups and/or electron withdrawing groups (e.g. nitro, cyano, halogeno etc.) which may further decrease the energy required for excitation giving absorption at longer wavelengths. This and other treatments of light absorption are useful when trying to predict the colour of new dye products by considering the molecular orbitals and their energies.

10.3   Classification of dyes and pigments

There are many ways of classifying species which absorb visible light [5,7,9]. Two which are frequently used are those by the Colour Index (acid, basic, direct dyes etc.), and by chemical structure (e.g. azo dyes, anthraquinones etc.). They are often used together and one can have, for example, direct azo dyes and also anthraquinone acid dyes.

10.3.1   *Colour Index classification*

10.3.1.1   *Acid dyes.* These dyes are of limited use in the paper industry. They contain one or more water-solubilising anionic groups which are almost invariably sulphonic acid groups or their sodium salts. In solution they dissociate to form anions of the dye and so are frequently called anionic dyes. Generally, the more sulphonic acid groups the greater the solubility, but affinity and/or substantivity for the substrate may be sacrificed if they become too soluble (i.e. poor bleed fastness or coloured backwaters may result). They are mainly applied to textile fibres having cationic character such as protein fibres (wool, silk, leather) or synthetic polyamides (nylon). These dyes have little substantivity and affinity for cellulose and, when applied to paper, they need fixing with cationic fixing agents or alum. The poor dyeing and fastness properties of acid dyes for paper is generally due to the fact that they are quite small, often non-planar molecules. Since cellulose is anionic in water, these dyes may therefore only be attracted by means of hydrogen bonding or through Van der Waals' forces (cf. section 10.5.1.2). For these forces to come into effect the dye molecules must approach very closely to the cellulose chains, for which linearity and/or planarity is required. They are quite bright dyes but for paper application are less suitable because they give rise to heavily coloured backwaters and are pH sensitive. However, they can be used satisfactorily by controlling the dyeing method. They are represented in nearly all the chemical dye classes, e.g. C.I. Acid Red 1 (**1** – Azo Geranine 2G) and C.I. Acid Blue 25 (**2** – an anthraquinone acid dye).

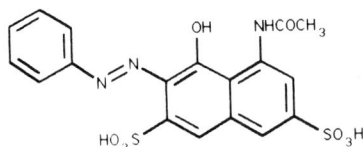

(**1**)

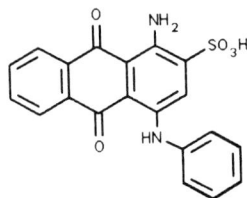

(**2**)

10.3.1.2　*Basic dyes.* These are cationic soluble salts of coloured bases. The positive charge, which may be delocalised over the whole of the molecule, is involved with the chromophore. These delocalised cationic dyes must not be confused with the pendant cationic dyes which have the positive charge located on one particular atom of the molecule isolated electronically from the chromophore (by methylene $-CH_2-$ bridges or similar). In these cases, the charged group has very little to do with the colour (cf. section 10.3.1.3). Being cationic, basic dyes are applied to substrates with anionic character where electrostatic attractions (salt links) may be formed. They have attraction for anionic groups present in protein fibres (wool, silk, leather), synthetic polyamides (nylon) and pulps containing lignin (e.g. ground wood or unbleached pulp). When acrylic fibres were developed (Acrilan, Orlon etc.), traditional basic dyes were applied, but soon, special basic dyes were developed for application to acrylics some of which have been adapted for paper application (cf. section 10.3.1.3.2). Basic dyes are not used on cotton as their structures are neither planar nor large enough for sufficient substantivity or affinity. On paper, which contains lignin and which is more anionic, the older basic dyes are applied (Bismark Brown, Auramines, Malachite etc.). On bleached pulp these dyes have very poor affinity and consequently are not generally suitable. Despite their poor lightfastness, they are used especially on packaging grades of paper and board. They are bright, with intense colours and are quite economical. They account for *c*. 30% of paper dye usage. Examples are C.I. Basic Blue 3 (**3** – Basic Pure Blue B) and C.I. Basic Green 1 (**4** – Brilliant Green). The structures are drawn with the cationic charge over one atom, but in reality it is delocalised over the molecule and it is more correct to write several resonance hybrids with the cationic charge in the various possible locations.

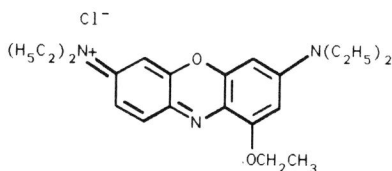

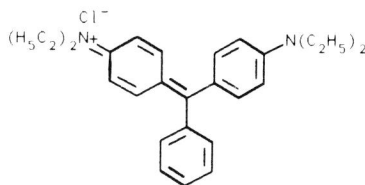

(3)   (4)

10.3.1.3   *Direct dyes.* Also called substantive dyes, these are the most important class as far as paper dyeing is concerned. Originally, direct dyes developed for dyeing cotton were used. Certain brands of these older dyes are still applied to paper, but now direct dyes specifically developed for paper are used. A direct dye must exhibit both good substantivity (the ability to adsorb onto the fibres from an aqueous medium) and affinity (the ability to become bound to the fibre).

10.3.1.3.1   *Anionic direct dyes.* These are chemically similar to acid dyes in having water solubilising anionic groups (almost exclusively $-SO_3H$), but they are different in having a greater substantivity and affinity for cellulose as they are larger, linear and/or planar molecules with extensive, delocalised π electron systems. In this way the dye molecules may approach the cellulose chains more closely and there is the possibility of hydrogen bonding, Van der Waals' forces and hydrophobic interactions coming into effect (in decreasing order of strength, cf. section 10.5.1.2.). These dyes have a good affinity for bleached cellulose fibres and are applicable over a wide pH range to give good all-round fastness properties. Often, cationic fixatives are necessary in conjunction with the dye to improve the retention and bleedfastness.

A good example of an anionic direct dye which was developed for paper application is described in [2,10,11]. C.I. Direct Red 81 (**5** – see Scheme 10.1 below) was originally applied to cotton, and when it was first applied to paper it showed good fastness but poor solubility (important if a liquid dye form is to be used). However, if modified by debenzoylation to give a slightly bluer red dye (**6**), the solubility is much improved but unfortunately the substantivity and affinity are adversely affected, giving rise to heavily coloured backwaters and poor bleedfastness properties. Substantivity and affinity of direct dyes may sometimes be adversely affected if too many ionic water-solubilising groups are present, as these increase the affinity of the dye for the aqueous phase, and the negative charges on the dye molecules actually have an electrostatic repulsion away from the negatively charged cellulose chains. Direct Red 81 can be improved for paper application by introducing more hydroxyl groups into the dye molecule which may form more hydrogen bonds with cellulose. The resultant dye, C.I. Direct Red 253 (7) has good solubility properties whilst still having good all-round dyeing properties (see Figure 10.3).

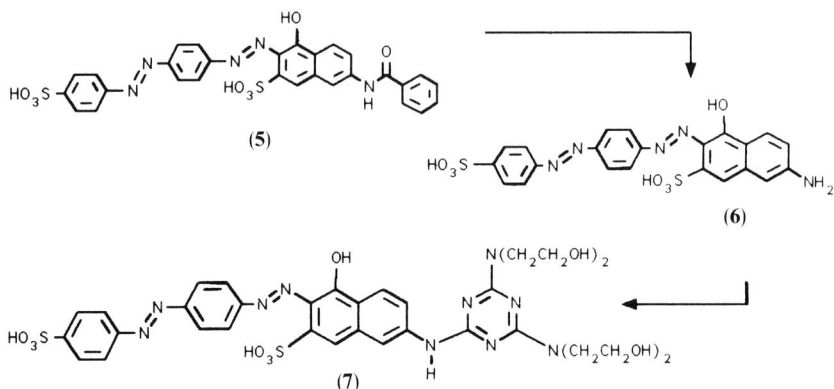

Scheme 10.1   Modification of Direct Red 81 for paper application.

Most anionic direct dyes (c. 70% of them) are polyazo dyes based on extended conjugation of double bonds. Their structures are based mainly on azobenzene, naphthalene, biphenyl, stilbene and aromatic heterocycles. Generally, longer wavelengths of light absorption may be achieved by extending the delocalised systems and by introducing electron donating and electron withdrawing groups into certain parts of the molecule. One possible way of improving the properties of a direct dye for paper application is to join two or more molecules together through the use of reactive bridging compounds. This greatly increases the molecular weight and, consequently, retention and bleedfastness properties are improved (assuming linearity and planarity are retained). Sometimes there is little or no change in colour. It is also possible to join different dye molecules together in this manner to produce mixed colours in one dye molecule. For example, a blue and a yellow dye (occasionally a poor combination for lightfastness due to catalytic fading) can be chemically bound together (with cyanuric chloride) to form lightfast green dyes such as C.I. Direct Green 26 (8). C.I. Direct Orange 118 (9) is an example of a dye whose molecular size is extended through the use of phosgene ($COCl_2$) which introduces a carbonyl group.

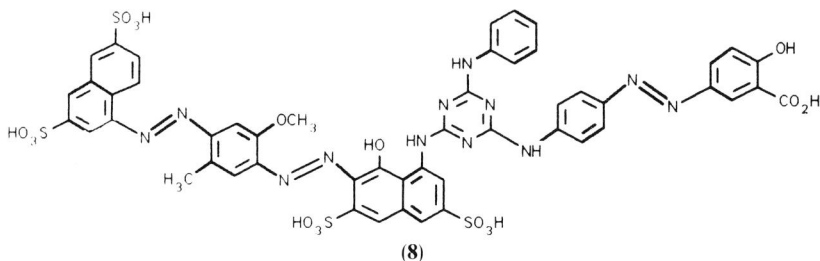

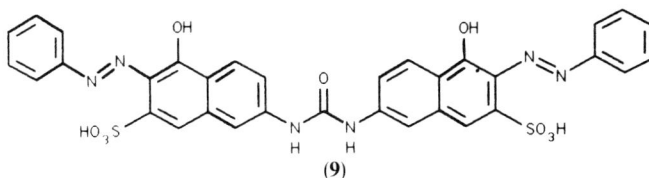

(9)

Fastness may also be improved by the use of metal complex dyes where, for example, Cu^{2+} complexes with certain parts of the dye molecule to give planar chelates. This is usually carried out before dyeing (premetallised dyes) since complex formation on the fibre can result in copper in the effluent. Other metals may sometimes be used (e.g. Cr, Fe, Co, Al, etc.) but for paper Cu(II) is almost exclusively used. If the complex formation involves the chromogen, as in *o*-hydroxyazo dyes such as C.I. Direct Blue 261 (**10**), then there may be a shade change, usually to more bathochromic (longer λ_{max}), often duller, colours. Complex formation may also occur with dyes containing salicylic acid residues (e.g. C.I. Mordant Orange 1 (**11**)) where little or no shade change is observed as the complexation occurs away from the main chromophore.

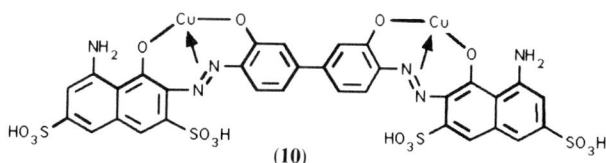

(10)

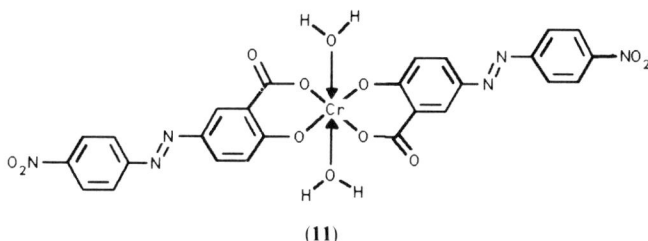

(11)

10.3.1.3.2 *Cationic direct dyes.* These are positively charged, and the charge may be pendant, residing over one particular location in the dye molecule (usually a quaternary nitrogen atom), or it may be delocalised. These dyes appear in the Colour Index as basic dyes, but this is strictly incorrect as they differ from the classical basic dyes in being larger and more linear and/or planar in much the same way as anionic direct dyes differ from acid dyes. Where there is a pendant cationic group attached to a direct dye molecule (e.g. $-CH_2CH_2N^+(CH_3)_3$), the substantivity for cellulose is greatly increased, so much so that paper can be dyed with virtually colourless backwaters even in cold water and with very short contact times (< 30 seconds). Due to the anionic character of cellulose in water, a cationic direct dye can form strong electrostatic attractions (salt links) in addition to the usual weaker H-bonds, and Van der Waals' and hydrophobic interactions. Thus Martin [2] reports how the dye based on modified C.I. Direct Red 81 (6) may also be treated to produce polycationic direct dyes such as Basic Red 111 (12). This

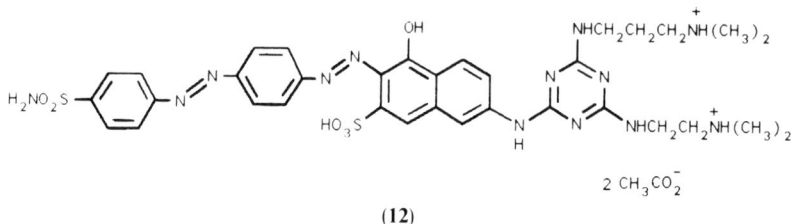

(12)

results in a greatly improved dye for paper application. The effect of such developments on the performance of dyes when applied to paper (better substantivity leading to improved retention) is illustrated in Figure 10.3. As seen in Figure 10.3, the red dyes developed specifically for paper application (7 and 12) show a great reduction in the amount of dye present in the backwater. The cationic direct dye performs the best, as even with very short contact times the backwaters are virtually free from dye (depending on depth of shade).

A similar improvement in dyeing properties has been noted by Arnold [11] and Martin [2], when considering a series of direct blue dyes (see Figure 10.4.).

As seen in Figure 10.4, the cationic direct dye (Cartasol Blue K-RL) achieves backwater and bleedfastness ratings significantly better than those achieved with the anionic ones (this is often, but not always, the case). There are many types of cationic groups besides quaternary ammonium groups. The positive charge may be partly delocalised over part of the molecule such as in systems containing pyridinium pyridone (13), thiazolium (14), benzthiazolium (15), indolium (16), and triazolium (17) cationic groups. These dyes are often made cationic by the methylation of heterocyclic nitrogen atoms.

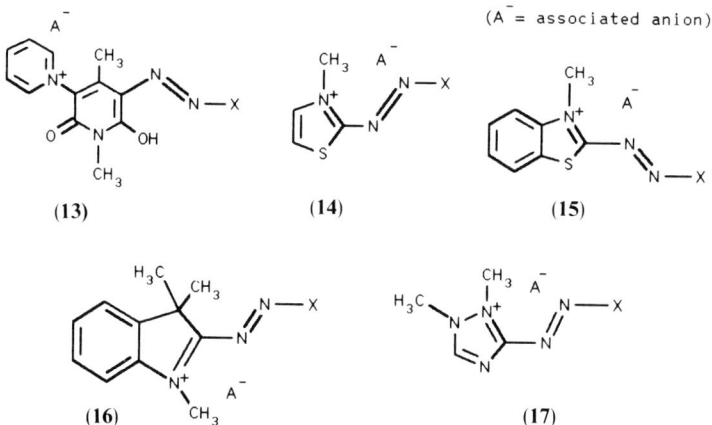

(13)   (14)   (15)

(16)   (17)

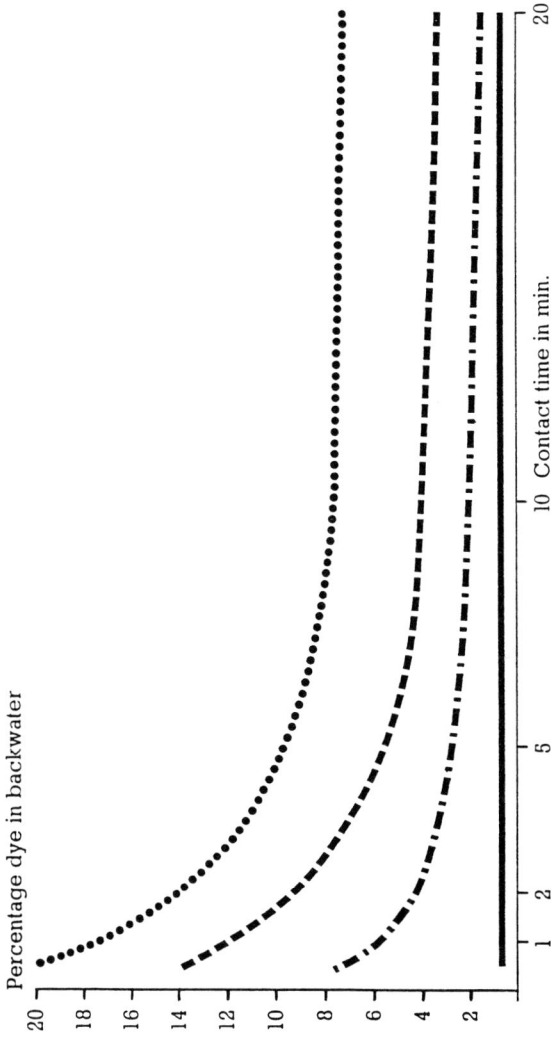

Figure 10.3 Improved performance of dyes based on Direct Red 81. (●●●● C.I. Direct Red 81 liquid, modified; ▬▬ C.I. Direct Red 81 powder/liquid; ▬ ▬ C.I. Direct Red 235 granulated/liquid; ▬▬ C.I. Basic Red 111 granulated/liquid). Fibre: tissue stock 20°SR; consistency: 25%; depth of shade: 1/6 standard depth.

Although the cationic charges are somewhat delocalised, they are not basic
dyes in the true sense as they are much larger linear and/or planar molecules
(e.g. **18**), making them more suitable for paper application. As with the anionic
direct dyes, the molecules can be made larger by using bridging groups such as
cyanuric chloride and urea, and they may also be complexed with metal ions if
they contain suitable residues.

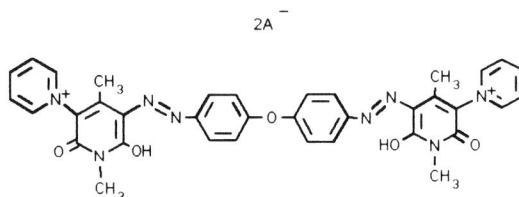

(18)

10.3.1.4 *Pigments.*   These colourants contain no solubilising groups and have
little, if any, affinity or substantivity for cellulose and therefore they require
fixing through the use of alum or fixatives. The fixation can be increased however by adding the pigment to the pulp prior to refining. An example is C.I.
Pigment Violet 23 (**19**). They are often applied as dispersions and there may be
an optimum particle size (more surface area results in stronger colour). They
are used extensively where high lightfastness is required, e.g. in inks where the
supporting medium attaches the pigment particles to the surface of the paper.
They are attractive to use for paper as they have excellent all round fastness
properties despite being expensive.

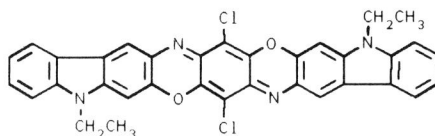

(19)

10.3.1.5 *Sulphur dyes.*   These are applied to paper only to a small extent. They
have quite large molecular sizes, are sometimes linear and/or planar but often
the structures are uncertain and may be mixtures. Sulphur dyes are used
mainly for black shades.

10.3.1.6 *Other dye classes not generally for paper application.*   Other classes
include reactive, vat and disperse dyes. Whilst sometimes applied to cotton
(except disperse dyes for synthetic fibres), they find little or no use on paper.
Suggested further reading is in [4,5,7,9].

10.3.2 *Classification by chemical structure*

10.3.2.1 *Azo dyes.* These are by far the largest and most important structural class, accounting for *c.* 70% of all dyes used. This is because they are tinctorially very strong, cover the whole shade range and are used in all applications as they are also very versatile. Their main disadvantage is that they tend to be a little dull especially in the blue shade region when compared with the next biggest class (anthraquinones). They contain azo linkages ($-N=N-$) which are planar and formed between a diazo component and a coupling component. The two lone pairs of electrons associated with the azo nitrogen atoms may delocalise fully into the π electron system and contribute to the colour of the molecule. The diazo component contains at least one free amino group which is capable of being diazotised (e.g. sulphanilic acid – **20**), which is then reacted with a coupling component (e.g. 2-naphthol – **21**). The diazotisation process can be carried out in various ways but most often by the use of hydrochloric acid and sodium nitrite (see Scheme 10.2). The resultant azo compound is C.I. Acid Orange 7 – 'Orange II' (**22**) used extensively to colour paper. It is inexpensive but requires fixing.

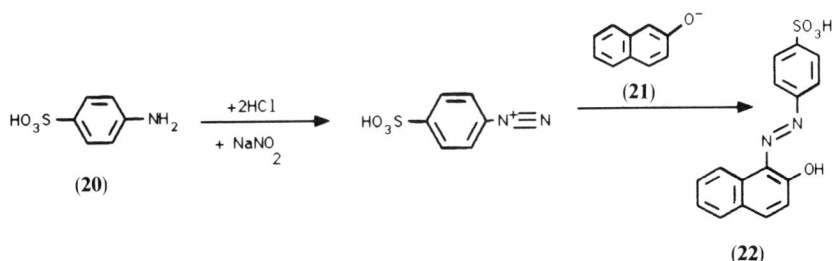

Scheme 10.2 Diazotisation of sulphanilic acid and coupling to 2-naphthol.

Through the use of diamino compounds for diazo components and/or coupling to compounds containing free amino groups, it is relatively easy to build up the large linear and/or planar molecules required for paper dyeing. Many different diazo and coupling components may be used. These are mostly derivatives of aromatic carbocyclic and heterocyclic compounds. Azo dyes are represented in all the classes of dyes (direct, acid, basic, disperse, reactive etc.).

10.3.2.2 *Anthraquinones and related carbonyl dyes and pigments.* Anthraquinone dyes are much weaker in colour and more restricted in their colour range than azo dyes. They find use mainly as acid, disperse or vat dyes and so are not normally applied to paper except as pigments. The molecules are generally too small to have much substantivity or affinity for paper (e.g. **23**). The related dyes also containing carbonyl groups ($-C=O$) include indigo (**24**) and many vat dyes based on condensed anthraquinones to form anellated compounds (e.g. **25**).

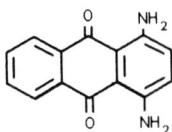

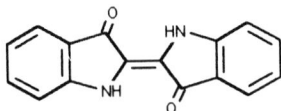

(23)   (24)   (25)

10.3.2.3 *Di- and tri-arylmethanes and related dyes and pigments.* This class forms the basis of most basic dyes (delocalised cationic type, Section 10.3.1.2). These are applied to wool, silk, acrylics, paper etc. They may also contain anionic groups such as in fluorescein (26) to become acid dyes. They are very bright and intense colours, often fluorescent due to their rigid structures (e.g. 27 – Rhodamine C.I. Basic Red 1), but often suffer from poor lightfastness properties. Related dyes include xanthenes (e.g. 28), oxazines (e.g. 29), thiazines (e.g. 30 – Methylene Blue C.I. Basic Blue 9) etc.

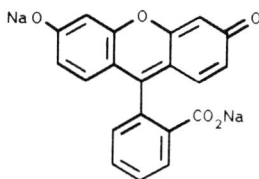

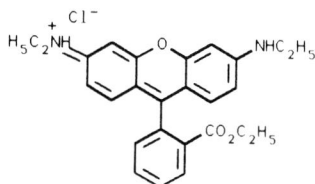

(26)   (27)

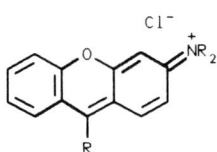

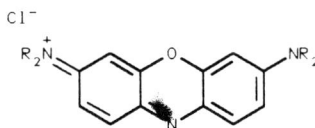

(28)   (29)

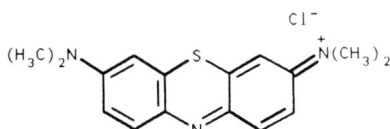

(30)

10.3.2.4 *Phthalocyanine dyes and pigments.* These are large planar metal complexes of phthalocyanine (usually Cu but sometimes Co, Ni, etc.). Copper phthalocyanine is a very insoluble, bright, turquoise pigment, but if sulphonated it may be applied as a direct dye for cotton and paper (C.I. Direct Blue 86 – 31) and it may also be chlorinated to various degrees to give greener shades. It may also be treated to introduce cationic groups (pendant type) and these are particularly suitable for paper (e.g. **32**). Similarly, phthalocyanines are found in reactive dyes.

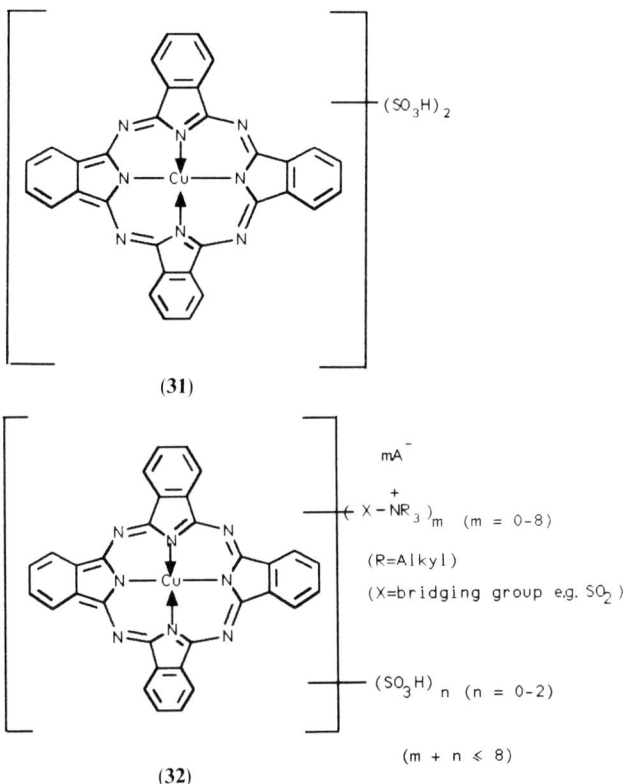

$(SO_3H)_2$

(31)

mA^-

$(X-NR_3)_m$ $(m = 0-8)$

$(R=Alkyl)$

$(X=bridging\ group\ e.g.\ SO_2)$

$(SO_3H)_n$ $(n = 0-2)$

$(m + n \leqslant 8)$

(32)

10.4 Dyes and pigments for paper

10.4.1 *Development of paper dyes*

Paper was originally coloured using inorganic pigments until textile dyes were used. The latter needed an affinity for paper fibres or to be fixed in the conditions present in the paper mill (which vary greatly). The obvious choice was anionic direct dyes, but these may need fixing to avoid coloured backwater. Basic dyes were found to be good for mechanical pulps since they are strong, bright and economical despite poor lightfastness. The market for paper dyes rapidly developed and now products are available to meet the requirements for paper application.

10.4.2 *Special requirements for paper dyes*

The requirements of a dye for paper are greatly dependent on the end use of the paper. Packaging grades of paper and board may be coloured economically with basic dyes where the fastness properties may not be important, whereas high quality writing paper or tissue require dyes with higher fastness properties.

10.4.2.1   *Physical forms.*   A review of the physical forms of dyes was recently made by Skelly [12]. Originally, powders were mostly used for dyeing textiles and paper, but recently dust hazards have made it more usual to obtain granulated dyes. These require dissolving, usually with heat, but cold-dissolving granules are becoming more popular. This dissolution is time-consuming and is only suitable for batch paper production and not for continuous processes. It is now more usual to use liquid forms of dyes as they are much easier to handle and can be metered directly into the papermaking process. The metering may even be computer controlled to regulate the shade (cf. section 10.5.3.5).

10.4.2.1.1   *Granules.*   The dye is usually made as a presscake or paste containing water and some electrolyte. This is then made into a solution or a homogeneous dispersion suitable for spray drying by the addition of the necessary agents. The spray drying conditions (temperature, nozzle width etc.) determine the granule size whilst the additives determine the wettability, solubility etc. The main advantage of granules over liquids is that they are much less bulky, giving more colour per kilogramme. Granules should be stable, dry, uniform, free flowing, odourless, easily dissolved (preferably in cold water), of high concentration (not too much electrolyte or other cutting materials), non-hygroscopic, non-dusting, non-foaming when agitated, and of high bulk density (granules are hollow and if too large may be bulky).

10.4.2.1.2   *Liquids.*   This form of dye is usually made as concentrated as possible from a presscake by increasing the dye solubility. This may be achieved by modifying the chemical nature of the dye, by ultra- or membrane filtration (to remove NaCl) or by the addition of water-miscible solvents and hydrotropic agents.

Both anionic and cationic dyes have a counterion present (usually Na^+ ions in the case of anionic dyes and Cl^- ions for cationic dyes), and this may be replaced by a different species to give greater solubility in water or water/solvent mixtures. With anionic dyes, Li^+ and NH_4^+ salts are more soluble than the Na^+ salt, and for cationic dyes, aliphatic acid ions (lactates, acetates or formates) often result in higher solubility. To further increase the solubility, solvents and hydrotropic agents are frequently added. For anionic dyes, solvents such as aliphatic alcohols, glycols, glycol ethers, alcohol amines, urea derivatives and many others are used. For cationic dyes, formic, acetic and lactic acids, aliphatic acid amides and many others are used in addition to those used for anionic dyes. The most commonly used hydrotropic agent is urea. The main advantage of liquids is their ease of use and time saving, but the main disadvantages are their bulkiness (*c.* 10–40% dye solutions) and occasional instability.

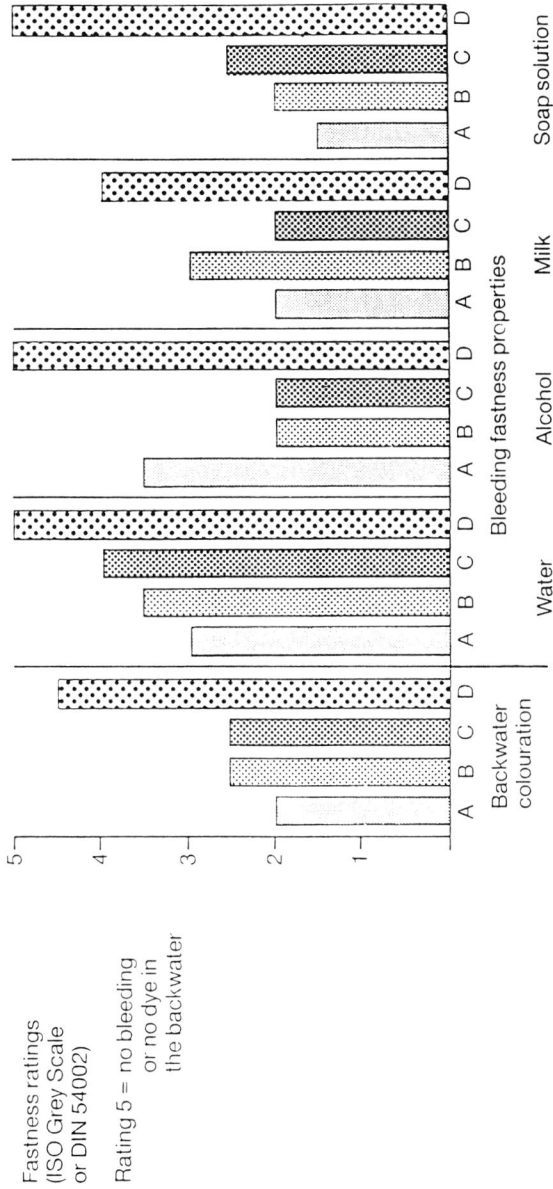

Figure 10.4 Backwater coloration and bleeding fastness properties of anionic and cationic blue dyes. Dyes A, C.I. Direct Blue 218/261; B, C.I. Direct Blue 1/10/15/281; C, C.I. Direct Blue 80/267; D, Cartasol Blue K-RL (cationic direct dye). Tissue stock 1/3 SD without fixative.

10.4.2.2 *Substantivity and affinity.* Dyes for paper should have high substantivity for paper fibres in order that as little as possible is wasted. Dyes with low affinity are also undesirable as fixatives and other additives may be needed to improve bleedfastness. The dye should preferably exhaust to completion in as short a time as possible (contact times as little as 20 seconds may be possible) and at very low temperatures (ambient temperatures in Scandinavian mills may be near zero). Cationic direct dyes have several features which give them superiority over their anionic counterparts due to increased substantivity and affinity:

- They give colourless backwaters and improved bleedfastness.
- They show fewer shade variations due to water hardness, pH (4–9) and temperature.
- They show excellent performance on unbleached, mechanical and waste paper pulps.
- They are insensitive to cationic auxiliaries present (no loss of colour yield due to precipitations as may occur with anionic direct dyes).

These generalisations depend on operating conditions and may not always be applicable.

10.4.2.3 *Stability.* It is very important that liquid dyes do not form sediments when left for long periods. They must be able to withstand many months of storage without forming deposits over a wide range of temperatures. If frozen, they must be able to thaw and form true solutions again without stirring. The viscosity of liquids is also important if they are to be pumped. Ideally, they must retain their low viscosity over a wide range of temperatures and this is especially difficult to achieve in the range over 0–5°C. Stability of granules (particularly in moisture content) requires them to withstand various temperatures and humidities without change. The dye should also be insensitive to the pH of the pulp and give the same stability over the range of pH encountered in paper mills (*c.* 4–9).

10.4.2.4 *Fastness properties.* The resultant dyed paper must pass a number of fastness tests dependent on end use. High quality dyed paper must not bleed out in water, soap solution, alcohol (and alcohol/water mixtures), milk, urine, acetic acid, etc., and be of sufficient fastness to light. Ideally, a zero-bleed dye is required and cationic direct dyes come closest to achieving this.

10.5 Application of dyes and pigments to paper

There are many different techniques for the application of colour to paper, *c.* 95% of dyes being applied to the stock and *c.* 5% at the size press. Conditions vary between machine, mill, end-use and location. There may be acid or neutral sizing systems, they may involve the use of different fillers, include other additives, and exhibit variations of temperature, water hardness, pH of incoming water etc. All of these may affect dyeing. The wrong method of application or even the wrong choice of dyes or addition point during the papermaking process can produce undesirable features in the finished product. Problems such as highly coloured backwaters, two-sidedness, mottling, poor bleedfastnesses, off shades etc. can be resolved by careful examination of the system being used to produce the paper and an understanding of the chemical nature of the paper, the colourant, other additives and the forces between them.

10.5.1   Theory of attractive forces between dyes and cellulose molecules

10.5.1.1   *Composition of wood.* The chemical composition of wood and the structure of cellulose is reviewed in great detail by Fengel and Wegener [13]. A knowledge of the chemical structures of the components of pulps helps in understanding the forces of attraction which are involved during the coloration of paper.

10.5.1.2   *Substantivity and affinity of dyes for paper.* The terms substantivity and affinity are different and are described by Wegman [14]. Substantivity of a paper dye is understood to be the ability of a dye to be adsorbed by cellulose fibres from an aqueous medium. The word affinity describes the capability of the dye to be bound to the fibre. Thus it is possible that a highly substantive dye applied with 100% exhaustion may show poor fastness due to a low affinity. Similarly, a dye of low substantivity and which gives coloured backwaters may however exhibit good fastness due to high affinity.

10.5.1.2.1   *Electrostatic forces.* When in water, cellulose is slightly anionic due to partly dissociated carboxylic acid (–COOH) and other groups (see Chapter 3). Some chemically treated pulps may also contain sulphonate groups. Therefore anionic dyes such as acid and anionic direct dyes experience a slight electrostatic repulsion during dyeing and hence rely upon other forces or require fixing. Cationic dyes such as basic and cationic direct dyes have an electrostatic attraction for these anionic groups but still require further forces or fixing. Some degree of spatial alignment between the cellulose and dye molecules is also required in order that the counter ions can be in close proximity for attraction to take place (easier for a cationic direct dye molecule than a basic dye).

10.5.1.2.2   *Hydrogen bonding.* Dyes containing hydroxyl, amino, azo, or amide groups (also many others) are capable of forming these bonds with the OH groups present in cellulose. These are much weaker forces of attraction than electrostatic forces and require the molecule to be oriented in a particular way in order that hydrogen bonding may occur.

10.5.1.2.3   *Van der Waals' forces.*   These forces are even weaker than hydro-gen bonding and require the molecules to be very close together before they come into effect. Direct dyes with their large, delocalised, linear and/or planar structures are able to approach the cellulose chains closer than other dyes, and these attractive forces therefore come into effect. It is thought that the large, delocalised π electron systems of the double bonds of direct dye molecules have an attraction for the positively charged nuclei of the atoms in the cellulose chains.

10.5.1.2.4   *Hydrophobic interactions.*   These are yet weaker forces and can be thought of as an attraction between the hydrophobic parts of a dye molecule and the hydrophobic parts of the cellulose chains. Cellulose is not as hydrophobic as some synthetic fibres such as nylon and polyester, so this is a very weak force indeed. If the dye molecule contains an alkyl chain within its structure, then it may exhibit a slightly greater affinity when the alkyl chain is longer.

10.5.2   *Influences of various factors used in the colouration of paper stock*

The extensive volumes edited by Casey [15] cover many of these areas of paper production.

10.5.2.1   *Influence of stock.*   Different types of pulp may give different results. There is firstly an inherent colour to the paper depending on the mechanical content (pale dull yellow). Also the degree of beating may give large variations in the shade since the beaten and fibrillated fibres are more accessible to dye molecules. Different dye classes have varying dyeing properties depending on the type of pulp.

(1) **Basic dyes** have low substantivity for bleached pulp and require fixing with anionic fixatives. They have a higher substantivity for mechanical pulps. Application to mixed bleached and unbleached fibres may therefore give mottling. Basic dyes are thus applied mainly to mechanical pulps for packaging and wrapping grades, chip board and cheap writing and printing papers.

(2) **Acid dyes** also have low substantivity for all pulps and require fixing with cationic fixatives. But even with fixing, the backwaters may be coloured. Therefore they are applied mainly to sized printing or writing papers and packaging papers. However, their use is declining.

(3) **Pigments** have virtually no substantivity for any pulp and only give appreciable shades through fixing. They are used in lightfast writing, printing, labels, accounting, document and laminating papers. They may exhibit two-sidedness when used in stock additions, but this may be unimportant where only one side of the paper is seen.

(4) **Anionic direct dyes** have high substantivity for bleached and unbleached

pulps occasionally giving only slightly coloured backwaters (depending on depth of colour). The bleedfastness is generally good but may improve on treatment with a fixative. It may even be possible to fix an anionic direct dye with a cationic direct dye in some situations giving shades of improved fastness. For pale to medium shades, no fixing agent is necessary. They are used for high quality writing, printing, tissue and decorative papers.

(5) **Cationic direct dyes** have excellent substantivity and may be used in all types of pulps to make all grades of paper (except for packaging grades where price may be the main consideration). They are superior to their anionic counterparts since they are much less sensitive to variations in pH, temperature, water hardness etc. (cf. section 10.4.2.2). Some cationic and anionic direct dyes may exhibit two-sidedness and problems may be overcome through the use of mixed cationic and anionic dyes. The cationic dye is applied first and then the anionic dye giving improved bleedfastness essential for hygienic papers.

10.5.2.2 *Influence of fillers.* Fillers are essentially cheap white pigments added to pulp to increase the opacity, whiteness, printability and other properties of paper (see Chapter 11). Most often used are china clay (kaolin), chalk ($CaCO_3$), talc and titanium dioxide. There are many grades of these fillers and sometimes they are used because they are cheaper than pulp (not titanium dioxide). For dyed paper, however, they will give weaker depths of colour. As fillers may be washed through the wire and as dyes have varying degrees of affinity for them, two-sidedness may become a problem. This is often remedied by the use of another dye, retention aids, or adjustments in the process (for example, the point or order of addition).

10.5.2.3 *Influence of sizing system and fixatives.* There are three main sizing systems, acid sizing (rosin and alum), neutral sizing (synthetic sizes) or surface sizing. The former two are discussed in Chapters 8 and 9. Size is added to pulps in order to modify the surface characteristics of paper by reducing water penetration. This is vital for writing and printing grades to reduce 'feathering'. Acid sizing (pH *c.* 3.8–6) may affect dyeing in certain cases and it is an undesirable feature of a dye to give different results with different systems. Different shades may be obtained by pre-sizing instead of post-sizing (sizing before or after dyeing) and the same applies to pre- and post-fixing. As long as the order of additions remains constant, only slight, acceptable differences in shade should result.

10.5.2.4 *Influence of water quality.* The quality of the incoming water to a paper mill may vary tremendously (especially if river water is used instead of town water) and the temperature, hardness, interfering substances present, pH etc. may all vary (sometimes seasonally). Although these changes may be gradual, frequent monitoring should be practised.

(1) The **temperature** of the water may be near freezing in Scandinavian mills and this may reduce the exhaustion of certain dyes.

(2) The **pH** of the water may vary between 4 and 9, therefore dyes developed for paper application should not be sensitive to pH variations within this range.

(3) The **water hardness** (due to dissolved Ca and Mg salts) may vary and certain anionic dyes may dye much weaker in soft water. This can be overcome by the addition of $CaCl_2$ or $MgSO_4$ to the water.

(4) **Interfering substances** such as humic acids from peat soil areas may be present. To a mill producing white paper, these substances may give brown discolorations, and for coloured paper they may cause loss in colour yield.

10.5.2.5 *Order of addition.* Variations in the method of applying dye, size, fixing agents and other additives may affect the resultant shade. Ideally, no variations should occur, but when they do, adjustments may be required in the order (or point) that the dye, size etc. are added. When applying dye mixtures, compatibility of the dyes may be a problem. When dyeing in stages, the order in which the dyes are applied may produce variations (e.g. cationics applied before anionics).

10.5.2.6 *Other additives.* Besides dye, size, fixatives and fillers, there may be other additives such as retention aids, dry- and wet-strength agents etc. Each may influence the results (e.g. certain anionic dyes may precipitate in the presence of cationic additives and produce coloured spots).

10.5.3 *Common problems and possible causes and cures*

10.5.3.1 *Coloured backwaters.* The dye may be the most expensive part of the paper process so it is essential that it is not wasted in the backwater. Mills recycle water wherever possible. Dye in the backwater may be due to variations in temperature, pH, water hardness or where fillers are used. Retention aids, fixing agents or a change in dye may cure this.

10.5.3.2 *Two-sidedness.* The final sheet of paper is not uniform as the shorter fibres (fines) and fillers partly wash through the wire until the longer fibres act as a filter. Therefore the top side (felt side) of the paper may contain more fillers and fines than the wire side. The extent of two-sidedness depends on the substantivity of the dye for both the fibres and the filler.

All dyes tend to dye fines a stronger colour and so the top side may appear darker. Alternatively, if the filler is undyed then the top side may appear weaker. Two-sidedness may be investigated by means of a Sandoz laboratory pulsator which can reproduce the same effect as the paper machine. This technique allows the two-sidedness to be reduced by adjustments in the recipe and/or point and order of additions.

10.5.3.3 *Poor bleedfastness.* The ideal is a zero-bleed paper dye which dyes the pulp fast to water, soap, milk, alcohol (alcohol/water), urine, acetic acid etc. The end use of the paper may allow the use of dye which is not zero bleed but fast to the conditions of that use.

10.5.3.4 *Unlevelness.* This may show as mottling or spots. Mottling occurs when there is poor circulation where the dye is added (such as in some batch dyeing processes) or when the dye is highly substantive and/or too concentrated. This should be overcome by dilution of the dye or better circulation of the stock, preferably both. Spots of colour may be due to precipitation of the dye. This is often due to pH variations or the formation of aggregates and/or interaction of counterionic species to form coloured ion pairs which are insoluble.

10.5.3.5 *Colour control.* Ideally, each shade should be produced using just one dye. Theoretically, a mixture of dyes could be used and most colours obtained from just four colours: red, blue, yellow and black. The state of the art is to use computer-controlled metering of dyes using systems such as the one illustrated in Figure 10.5. Liquid dyes are stored in tanks and pumped automatically as required into the stock (dilution is usually necessary). Such dyeing systems are necessary where the colour of the finished paper is measured by a colorimeter and adjustments automatically made to the flow rates of the liquid dyes if the colour varies slightly from that required [16, 17].

Figure 10.5 A 10 colour automatic metering system.

10.5.4 *Printing of paper*

Printing is a surface treatment for paper and the colours used in the inks are not required to have any special structural conformities such as linearity or very large delocalised π electron systems. The colour is usually in the form of a pigment in a medium which binds to the paper surface. All colours may be obtained using only four inks: red, blue, yellow and black (in fact magenta, cyan, yellow and black) which may be printed in dots in various processes (e.g. photogravure, lithography).

10.6   Fluorescent whitening agents (FWAs)

Also called fluorescent brightening agents (FBAs) or optical brightening agents (OBAs), these compounds are applied to paper as well as to textiles. In 1984 Zollinger [9] estimated that some 33 000 tonnes of FWAs (active material) were produced. About 50% are put into detergents, about 33% in paper, 15% in textiles and 3% in plastics. 80% of FWAs are based on stilbene compounds (see below) and they may be considered chemically similar to anionic direct dyes since they are also quite large planar/linear molecules with large delocalised π electron systems and one or more sulphonic acid ($-SO_3H$) groups.

10.6.1 *Concepts of fluorescence*

Detailed accounts of fluorescence and related topics are found elsewhere [7, 9]. The mechanism of fluorescence is similar to that for coloured dyes (see section 10.2.2), but in the case of FWAs, the energy difference ΔE is much larger and consequently the incident radiation must be of shorter wavelength for excitation to occur. The incident radiation required for excitation is in the near ultraviolet region of the electromagnetic spectrum (*c.* 240–380 nm). In a dye molecule, the absorbed energy may be dissipated through radiationless transitions such as vibrational relaxations, but when the structure of the molecule becomes more rigid such transitions become more restricted. If the structure is sufficiently rigid, the molecule may emit radiation as fluorescence in order to lose the energy. Fluorescence usually occurs from the lowest energy excited state to the ground state and the fluorescence emission spectra are always of longer wavelengths than the absorption spectra (see Figure 10.2). In the case of FWAs, the absorption occurs in the near UV region and emission occurs at short visible wavelengths (*c.* 400–500 nm), giving additional violet–blue light to the reflected light. Thus a white sheet of paper made from bleached pulp may be treated with an FWA and appear whiter. The inherent yellowness of paper pulps (even after bleaching) may be removed by the use of white pigment fillers (TiO_2) and a trace of a blue–violet dye. This tinting at very low levels (sometimes 0.005% or less) gives the appearance of a whiter sheet. A much brighter white may be achieved by incorporating an FWA and this gives white paper with improved contrast for writing and printing paper.

The amount of FWA and the method of application can vary in much the same manner as dyes. The amounts, however, are generally quite low ($c. \leqslant 0.2\%$ as active material) for sufficient brightening. If more is used it is possible that the whiteness increases up to a certain extent as the FWA builds up, but then the whiteness decreases to give what is often called greening or yellowing. This is due to the inherent colour of the FWA applied since, at higher concentrations, the FWA may absorb some visible light in the blue region. FWAs are applied by addition to the stock or as a surface treatment in a size press or coater, often a combination of the two for higher whites (cf. section 10.6.4).

10.6.2 *Chemical constitution of FWAs*

Various books and reviews are suggested for a more detailed account of FWA synthesis and chemical types [5, 9, 18, 19].

10.6.2.1 *Stilbenes.* FWAs based on stilbenes (one or two stilbene residues) account for *c.* 80% of commercially available brighteners. Most are based on 4,4'-diaminostilbene-2,2'-disulphonic acid (**33**) and may be obtained by reaction with cyanuric chloride (**34**) and condensation of the product (**35**) with suitable compounds to introduce groups such as amino, alkylamino, arylamino, hydroxy, alkoxy, aryloxy etc. to give a symmetrical, planar, fluorescent molecule (**36**) (Scheme 10.3).

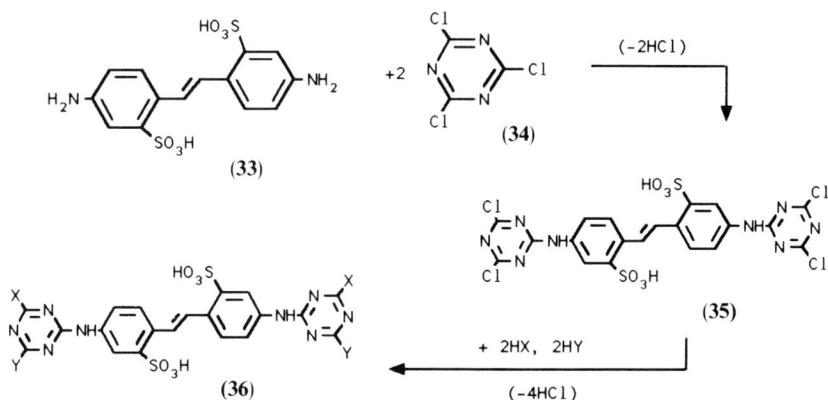

Scheme 10.3 Stilbene based FWA synthesis.

Unsymmetrical FWAs based on **33** are possible, as well as those based on systems with two stilbene residues to give larger molecules with extended conjugation (e.g. **37**).

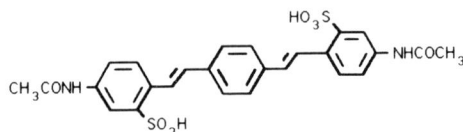

(**37**)

10.6.2.2 *Other FWAs.* There are many other chemical types but they are generally not often applied to paper but to textiles (paper brighteners are almost exclusively stilbene based). They are mainly coumarin derivatives (e.g. **38**), 1,2-ethylene derivatives (e.g. **39**), diarylpyrazolines (e.g. **40**), and naphthalimides (e.g. **41**).

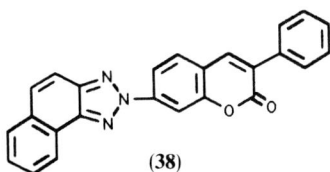

(**38**)

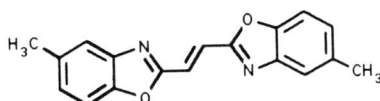

(**39**)

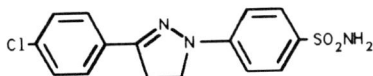

(**40**)

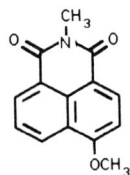

(**41**)

10.6.3 *Structure and fluorescence*

The chemical constitution of an FWA molecule depicts its properties. For FWAs, there is a measurement of the efficiency of fluorescence, the quantum yield Φ_f, which is as important as the UV light absorption. The quantum yield Φ_f is a measure of the number of FWA molecules which, once excited, lose their energy to the ground state by fluorescence and not by other means (vibrational relaxations etc.). If all excited molecules fluoresce, then Φ_f is 1.0. If only 50% of excited molecules fluoresce then Φ_f is 0.5. Electron withdrawing groups present in the chromophore (e.g. $-SO_3H$) may decrease the fluorescence (smaller Φ_f), whereas electron donating groups (e.g. $-OH$, $-NH_2$, $-OCH_3$) may increase it. However, the substrate is also important and this may not always be the general situation. Also, electron donating groups may increase the wavelengths of fluorescence and UV absorption (bathochromic shift) and electron withdrawing groups may decrease them (hypsochromic shift).

The substituents present in the FWA chromophore may also affect the solubility, substantivity and affinity for cellulose. As with direct dyes, sulphonic acid groups are present to confer solubility but if too soluble there may be a detrimental effect on the substantivity. It is more favourable to have as few sulphonic acid groups as possible (one or two) and then the solubility of the FWA molecule may be improved by the introduction of groups such as amino, alkylamino, hydroxy, alkoxy etc. with the affinity or substantivity being enhanced rather than adversely affected.

10.6.4   *Application of FWAs to paper*

FWAs are not generally applied to mechanical pulps as the lignin quenches the fluorescence (lignin also absorbs UV light). Similarly, FWAs applied to TiO_2-filled papers will have reduced brightness since TiO_2 also absorbs UV light. Certain chalk and clay fillers may reduce brightness since they may counteract the fluorescence by being slightly yellow and dull.

FWAs may be applied directly to the stock in the same manner as one would use a direct dye, or they may be applied on the surface of the paper at the size press or by coating. Both sized and unsized grades of paper may be brightened, but the maximum whiteness is obtained in neutral or slightly alkaline conditions.

10.6.4.1   *Requirements of FWAs for paper applications.*   The substantivity of paper brighteners applied to the stock must ideally be high, but FWAs which are applied at the size press can have lower substantivity. Universal FWAs may be applied either to the stock (any sizing system) or at the size press. There are other FWAs which are more suitable for acid sizing systems (acid stability required), neutral sizing systems, or surface application (size press or coating systems).

Figures 10.6 and 10.7 illustrate the application of three FWA products. Leucophor AP is a high substantivity FWA (possessing two $-SO_3H$ groups) most suitable for application in the stock, Leucophor U is a medium substantivity FWA (four $-SO_3H$ groups) for universal application methods, and Leucophor SHR is a low substantivity FWA (six $-SO_3H$ groups) most suitable for application at the size press.

In Figure 10.6 the highly substantive product Leucophor AP has low concentrations of FWA in the backwaters with quite short contact times and it is much less sensitive to the water hardness than the universal type of FWA Leucophor U (medium substantivity). For this latter product, very soft water should be avoided or artificial hardening with $MgSO_4$ or $CaCl_2$ should be practised.

Figure 10.7 shows how the three FWAs build up when applied to paper by a size press. Here, the highly substantive FWA is less suitable, building up to a relatively low degree of whiteness, whereas the universal FWA and the low substantivity FWAs build up to higher degrees of whiteness.

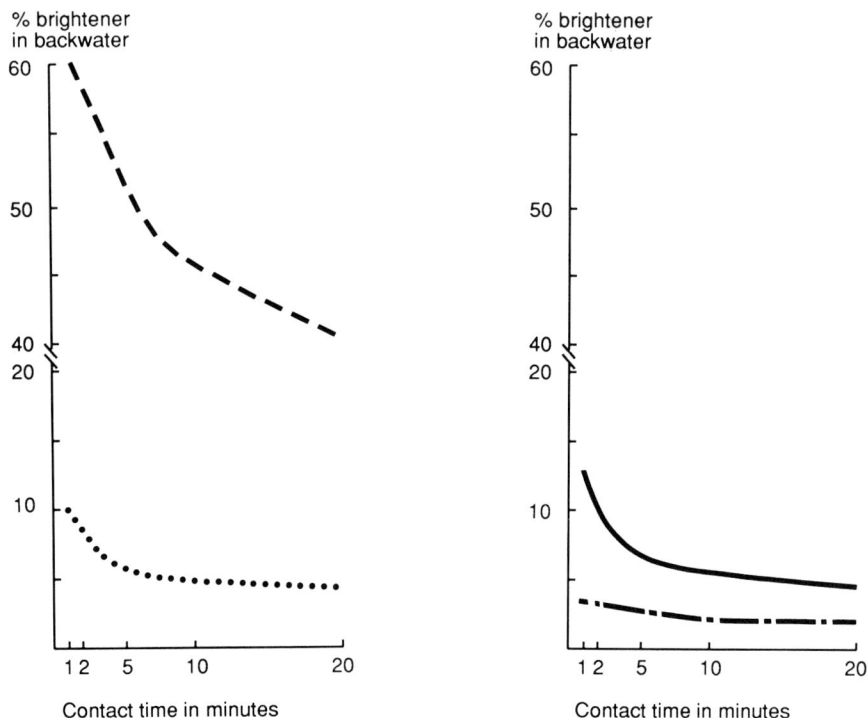

Figure 10.6 Brightener in the backwater as a function of the water hardness and the brightener/pulp contact time. Brightener applied: – – – 0.4% Leucophor U liquid at 1° dH/ 1.25° Bh; ··· 0.4% Leucophor U liquid at 12° dH/15° Bh; — 0.4% Leucophor AP liquid at 1° dH/1.25° Bh. ·—·—· 0.4% Leucophor AP liquid at 12° dH/15° Bh. Stock: tissue pulp 20° SR.

10.6.4.2 *Stability of FWAs.* As with direct dyes, FWAs may be supplied as powders (preferably non-dusting, cold-dissolving granules) or as solutions. These physical forms should ideally exhibit similar stabilities (cf. section 10.4.2). Their stability to light however is usually inferior to that of dyes so they photodegrade more rapidly which can lead to the gradual build up of yellow photodecomposition products. This is partly due to the much longer-lived excited states of FWA molecules in comparison with dyes where photochemical reactions (usually with atmospheric oxygen) can occur. With dyes, the light absorption and vibrational relaxation processes occur very quickly (first order rate constants $10^{15} \sec^{-1}$ and $> 10^{12} \sec^{-1}$ respectively). An FWA molecule, however, takes much longer to lose its energy after absorption of UV light (first order rate constants for fluorescence are between $10^6 \sec^{-1}$ and $10^9 \sec^{-1}$). Liquid forms may be especially unstable and therefore require protection from strong light sources. Generally, electron withdrawing groups present in the chromophore may accelerate photodegradation whereas electron donating groups may retard it.

Berger Degrees
of Whiteness

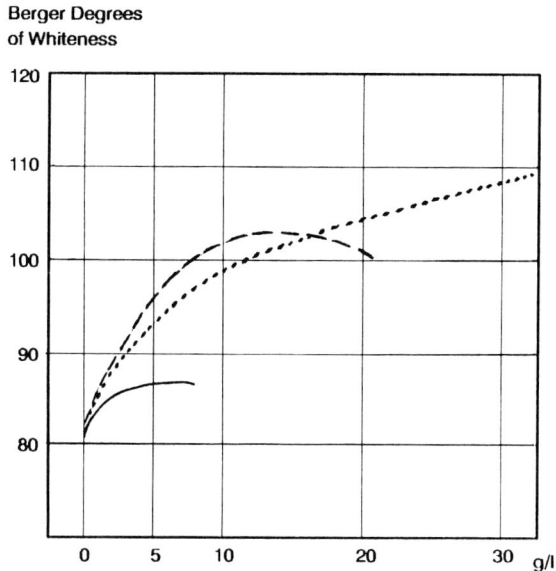

Figure 10.7   Build up of the Leucophor products on the size press. (... Leucophor SHR liquid; ---Leucophor U liquid; —Leucophor AP liquid). Size press solution: 50 g/l Farinex TSC; base paper: sized paper, pH 4.8, 15 g/m² pickup.

10.6.4.3   *Evaluation of whiteness.*   There are many formulae which try to give a basis for the evaluation of colour and whiteness through the use of colour measuring devices. A method of evaluation for the degree of whiteness which was agreed upon in 1983 at an international level within the CIE (Commission International de l'Eclairage) enables the calculation of a colorimetric value for the degree of whiteness and one for the tint. If Y = the Y tristimulus value and x and y are the x,y chromaticity coordinates, then:

CIE degree of whiteness: $W = Y + 800.(0.3138 - x) + 1700.(0.3310 - y)$

CIE tint value: $T = 900.(0.3138 - x) - 650.(0.3310 - y)$

This is now an ISO Standard technique (105-J02) using a standard type of light source D65 and the 10° normal observer. Other older degree of whiteness formulae are still in use however and they include the Berger, Taube, Hunter and Stensby formulae. Pure fluorescence-measuring devices are also used where a brightened sample may be irradiated with UV light only and the fluorescence in the visible region (not reflected UV light) is measured. This is only of use where a brightener is evaluated with reference to a standard.

10.7   Ecotoxicology

New research into the development of paper additives of all kinds is increasingly being driven by ecotoxicological concerns. The present day chemical producer must aim not only to minimise or eliminate any toxicological properties inherent to the additive, but also to prevent any release of the additive into the environment. Although the chemical constitution of the additive is of prime importance in this regard, the physical form of the additive must also be taken into consideration. Further discussion of this subject can be found in [20–23].

10.7.1   *Physical form*

10.7.1.1   *Solids.* Powders can be particularly hazardous because there is the potential for inhalation of airborne dust, contamination of the environment, as well as a risk of explosion (often encountered with dusts of various nature, e.g. sawdust, flour). Although solid forms of dyes and FWAs are less frequently encountered in the paper industry today, they retain some importance for certain uses and are typically made non-dusting. This can be accomplished either by adding an anti-dusting agent, e.g. during grinding, or by spray-drying a liquid formulation and producing a non-dusting granular form.

Solid forms generally require the presence of additional materials. Typically, these are either inorganic salts, dispersants or glucose, which are present to enhance solubility; to aid the dyeing process; and to standardise the concentration of the product. Although these materials may be harmless, they are not substantive to paper, and therefore are lost to the backwater where they may be regarded as having some environmental impact.

10.7.1.2   *Liquids.* It is now usual for the paper manufacturer to use dyes and FWAs in concentrated liquid form, but preparing this form is far from straightforward. The presence of inorganic salts left over from the production process tends to cause the dye or FWA to come out of solution and, in the past, solubilising auxiliaries have been added in order to prepare concentrated liquids. Such auxiliaries have included urea, various alcohols, amides and organic acids which inevitably accumulate in the backwater. Dyestuff producers are increasingly making efforts to reduce or completely avoid the use of such auxiliaries, with the target of developing liquid formulations containing only active product and water. One development has been to use ultrafiltration, or reverse osmosis, to reduce the salt content and the need for auxiliaries.

10.7.2   *Metals*

The use of metals, particularly heavy metals is avoided whenever possible, but in paper dyeing, metals are much less common than in textile dyeing, and their use is almost exclusively restricted to copper-complexed dyes to provide specific colouring or dyeing properties; in particular, lightfastness. The fact that the metal is chemically-bonded to the dye as a part of the production process means that there need be no free, biologically-available metal in the dye as supplied to the paper manufacturer.

Examples of copper-complexed dyes include copper phthalocyanines and C.I. Direct Blue 261 (**10**).

10.7.3   Chemical constitution

The design and development of dyes and FWAs with high substantivity and affinity to paper have been described earlier in this chapter, improvements in both these properties are continually being sought and achieved by the dyestuff producers, with obvious environmental benefits. Much emphasis is now being placed on the development of bleedfast dyes and FWAs; that is to say, dyes and FWAs which are not removed from the paper on contact with common solvents such as water, soap solution, acetic acid, urine or alcohols. This is of particular importance for hygienic papers and food wrappings. The possibility, however, that some dye or FWA can be released from the paper can never be ignored, and the utmost importance is given to rigorous toxicological testing of all new products.

An area of increasing concern is that of the mutagenic potential of dyes in general, and azo dyes in particular. It has been the case in the past that certain commercial azo dyes, while not necessarily mutagenic *per se*, have been shown to generate mutagenic aromatic amines through metabolic reductive cleavage of the azo linkages. C.I. Direct Blue 15, for example, is converted through metabolism into the known mutagen and carcinogen dianisidine (Scheme 10.4). This dye, once widely used, has now been withdrawn by most manufacturers from their product ranges.

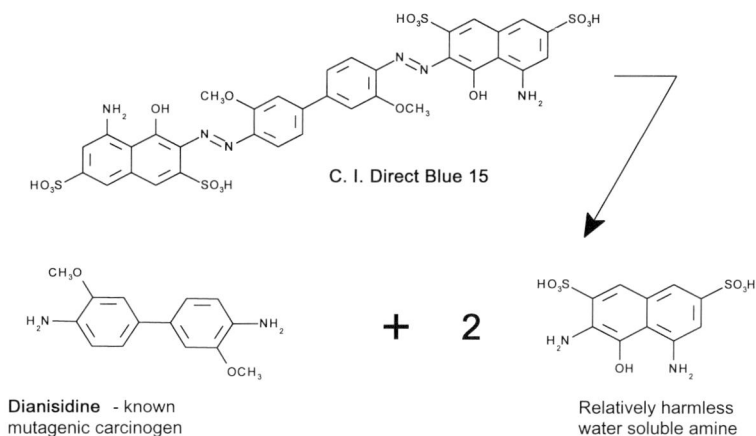

C. I. Direct Blue 15

Dianisidine - known
mutagenic carcinogen

+ 2

Relatively harmless
water soluble amine

Scheme 10.4   Fragmentation of C.I. Direct Blue 15 by metabolic reduction.

For this reason it is now becoming more common, when screening the mutagenic potential of a new azo dye, to supplement the standard *Ames test* for mutagenicity with the so-called *Prival* modification. This modification incorporates a reductive metabolic activation step, which promotes reductive cleavage of the azo linkages and therefore focusses on the safety of the constituent aromatic amines. Furthermore, an application to register a new dyestuff that could *in theory* be broken down to form products of known or suspected mutagenicity or carcinogenicity may now need to be accompanied by the results of long-term animal feeding studies in addition to the standard mutagenicity-screening test results.

Considerable research effort continued to be directed towards removing all potential for mutagenicity in azo dyes. One approach has been to construct the azo dye so that each aromatic amine contains at least one water-solubilising group; there is considerable evidence that hydrophilic amines generally have no mutagenic activity. A second approach has been to hinder the metabolic reduction process by attaching large chemical groups close to the site of the azo linkage. Replacement of the methoxy (OCH_3) groups in C.I. Direct Blue 15 with more bulky butoxy ($OCH_2CH_2CH_2CH_3$) groups for example, produces a dye with reduced mutagenic activity because the possibility of metabolic reduction is lessened (Hoechst, DE 3 511 545). Other approaches have been discussed by Freeman [23].

The design of ecotoxicologically more friendly dyes and FWAs, while maintaining as far as possible those properties which render the product attractive to the consumer, presents a considerable challenge to the chemical producer. A greater challenge is to achieve these product improvements without adding too much to the cost.

Acknowledgements

The author wishes to acknowledge the contribution of the following to this chapter: Dr J. Coates, Mr F. Colling, Dr N. Dunlop-Jones, Mr J. Farrar, Mr G. Martin, Dr H. Moser, Mr A. Tindal and Dr A.C. Jackson.

References

1. Perkin, W., British Patent 1984 (1856).
2. Martin, G., *Wochbl. Papierfabr. 110*, **13** (1982), 457.
3. Billmeyer, F.W. (Jr.) and Saltzman, M., *Principles of Colour Technology*, 2nd edn, Wiley Interscience, New York (1981).
4. Abrahart, E.N., *Dyes and Their Intermediates*, Arnold (1977).
5. Allen, R.L.M., *Colour Chemistry*, Thomas Nelson & Sons, London (1971).
6. Evans, N.A. and Stapleton, I.W., in *The Chemistry of Synthetic Dye*, ed. Venkataraman, K., Vol. VIII, Academic Press, New York (1978), 221.
7. Griffiths, J., *Colour and Constitution of Organic Molecules*, Academic Press, London (1976).

8. Straley, in *The Chemistry of Synthetic Dyes*, ed. Venkataraman, K., Vol. III, Academic Press, New York (1970), 385.
9. Zollinger, H., *Colour Chemistry – Syntheses, Properties and Applications of Organic Dyes and Pigments*, VCH, Weinheim (Germany) (1987).
10. Groebke, W. and Martin, G., *Rev. Prog. Coloration*, **14** (1984), 132.
11. Arnold, E. and Martin, G., *Paper Southern Africa*, May–June (1986).
12. Skelly, J.K., *J. Soc. Dyers Colour.*, **60** (1980), 618.
13. Fengel, D. and Wegener, G., *Wood – Chemistry, Ultrastructure and Reactions*, De Gruyter, Berlin (1983).
14. Wegmann, J., *Textil Rundschau*, **11** (1959), 631–642.
15. Casey, J.P., ed. *Pulp and Paper – Chemistry and Chemical Technology*, 3rd edn, Vol. III, Wiley Interscience, New York (1981).
16. Bartram, E., *Tappi*, **70**(7), July (1987), 71–72.
17. Lorditch, G.M., *Paper Technology and Industry*, **28**(10) (1987), 634–635.
18. Siegrist, A.E., Hefti, H., Meyer, H.R. and Schmidt, E., *Rev. Prog. Coloration*, **17** (1987), 39.
19. Sarkar, A.K., *Fluorescent Whitening Agents*, Merrow Publishing, England (1971).
20. Gregory, P. *High-Technology Applications of Organic Colorants*, Topics in Applied Chemistry, Plenum Press, New York, pp. 255–272.
21. Hunger, K. and Jung, R., *Chimia*, **45** (1991), 297–300.
22. Turner, P. *J. Soc. Leather Technol. Chem.*, **78** (1), (1994), 8–11.
23. Freeman *et al.*, in *Colour Chemistry*, eds Peters, A.T. and Freeman, H.S., Advances in Colour Chemistry Series, Elsevier Science Publishers, Barking, Essex (1991), pp. 85–114.

11 Physical and chemical aspects of the use of fillers in paper

R. BOWN

11.1 Introduction

Fillers have always been important raw materials in papermaking, and they are now used for various reasons in a wide range of paper and board products. The spectrum ranges from grades of newsprint containing as little as 3 wt% of a special filler, added to improve opacity and printing properties, to office and magazine papers containing 30 wt% or more filler, added partly to improve some paper properties but principally to reduce total raw material costs. Given the variety, it is beyond the scope of this chapter to discuss any individual fillers or applications in detail. Instead an attempt is made to identify the major chemical and physical principles that dominate the effect of all fillers both in the formation of paper and in the properties of the finished sheet. From these principles, a more general understanding of the current use of fillers and the opportunities for future development should emerge.

Fillers are used mainly for two simple reasons: they improve certain properties, such as brightness, opacity, smoothness and printing properties and if they are cheap, they reduce production costs. Fillers may then be classified into two broad groups:

(1) The general purpose fillers – those used at loading levels of greater than 10 wt% in the sheet. For these fillers there is a compromise between the economic aspects and the expectations with respect to improvements in sheet properties per unit weight of filler used.
(2) The speciality fillers – those used at loading levels generally less than 10 wt% and often less than 5 wt% in the sheet. For these fillers there is more emphasis on significant improvements in specific sheet properties, in particular optical properties or print related properties.

Typical examples of general purpose fillers are those derived from natural white minerals such as kaolin, chalk, limestone and talc, and characterised by a particle size distribution falling substantially within the range 10 μm–0.5 μm e.s.d.[1]. Examples of specialty fillers are titanium dioxide, calcined clay, urea formaldehyde, synthetic alumino-silicates and precipitated silica. Speciality fillers often have carefully controlled particle size distributions, high bright-

---

[1] Equivalent spherical diameter obtained from the rate of sedimentation in water and Stokes' Law.

ness and specific attributes such as high refractive index (e.g. titanium dioxide), low density (e.g. organic pigments) or aggregated particles containing considerable void volume (e.g. calcined clays, silicas, alumino–silicates). These attributes are obviously essential to improve required paper properties. Some fillers however, notably precipitated calcium carbonate, may be produced in a range of product forms and fall into either grouping depending on local usage and production economics. The general properties, availability and use of the various fillers are described elsewhere [1]. The more recent use of 'on-site' precipitated calcium carbonate is now also well described in the literature [2].

From the point of view of a fundamental study of papermaking, the most important properties of any filler are the particle size and shape (including any permanent aggregation or pore structure), specific gravity and surface chemistry. The discussion here deals mainly with the influence of these properties. Other bulk properties such as refractive index and light absorption are discussed where appropriate.

11.2   Filler properties

11.2.1   *Particle size and shape*

In a composite structure, such as paper, the relative sizes, shapes, number concentrations and surface contact areas of the various constituents can be very important, so that in a discussion of the role of a filler, knowledge of the filler particle size and shape distributions and their relationship to surface area and particle numbers per unit mass is essential.

There are a number of techniques commonly used to measure particle size distribution and these are based on measurement of phenomena such as translatory or rotary diffusion, sedimentation, optical scattering power and conductivity changes [3]. In each case it is common to interpret experimental data in terms of the theory applicable to spherical particles. The results are then quoted in terms of 'equivalent spherical diameters' (e.s.d.). However, as may be seen from the electron micrographs in Figures 11.1 and 11.2, paper fillers rarely contain spherical particles – indeed anisometric, platey and sometimes aggregated particles are commonly found. Therefore it is important to understand how measurements of e.s.d. relate to actual particle dimensions, and hence to surface areas and particle numbers. The relationship depends on the method of measurement, and since for paper fillers the most common form of particle size measurement is sedimentation in water, only this method will be considered here.

Stokes' Law is used to calculate e.s.d. from sedimentation measurements, and in its simplest form may be written as:

$$d = \left(\frac{18\, u\eta}{g(\rho - \rho_s)} \right)^{1/2}$$

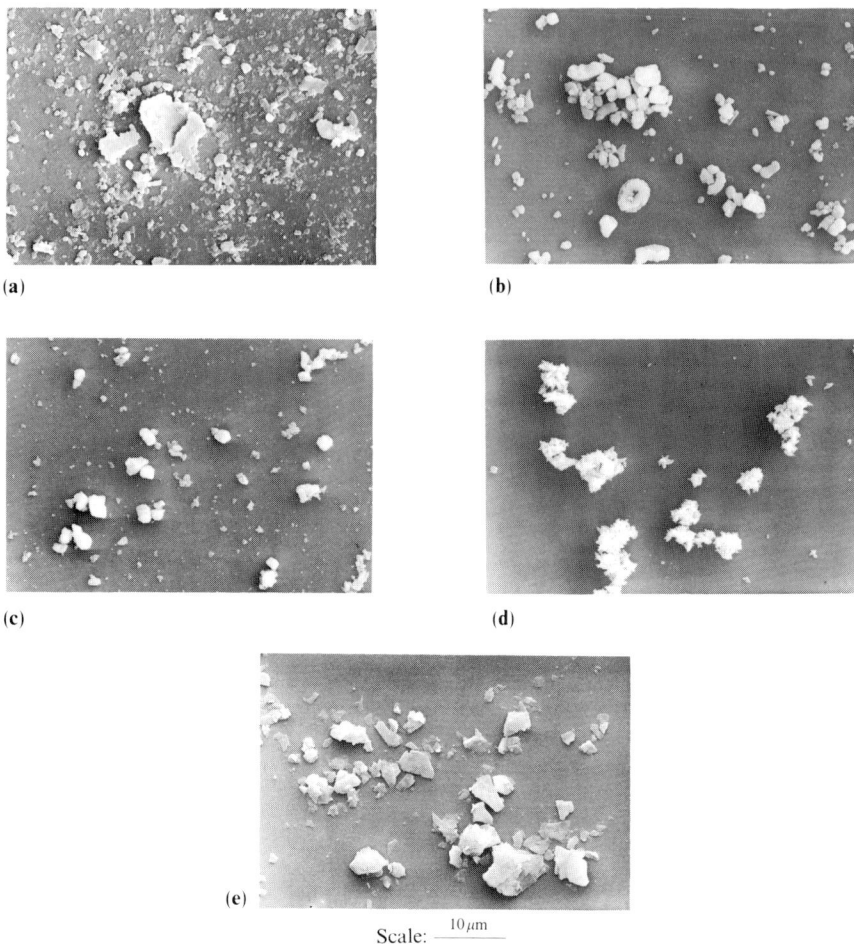

(a)

(b)

(c)

(d)

(e)

Scale: $\dfrac{10\,\mu m}{}$

Figure 11.1 Electron micrographs of some common fillers : (a) standard kaolin; (b) standard chalk; (c) ground limestone; (d) scalenohedral pcc; (e) talc.

where d is the e.s.d.

 η is the viscosity of the solution

 ρ and ρ_s are the specific gravities of the particle and the solution respectively

 u is the sedimentation velocity, and

 g is the acceleration (due to gravity or centrifugal force depending on the instrumental method).

There are two problems: the relationship between d and the actual dimension of anisometric particles and the apparent value of ρ for aggregated or porous particles.

In practical terms, fillers may be described as either blocky or approxi-

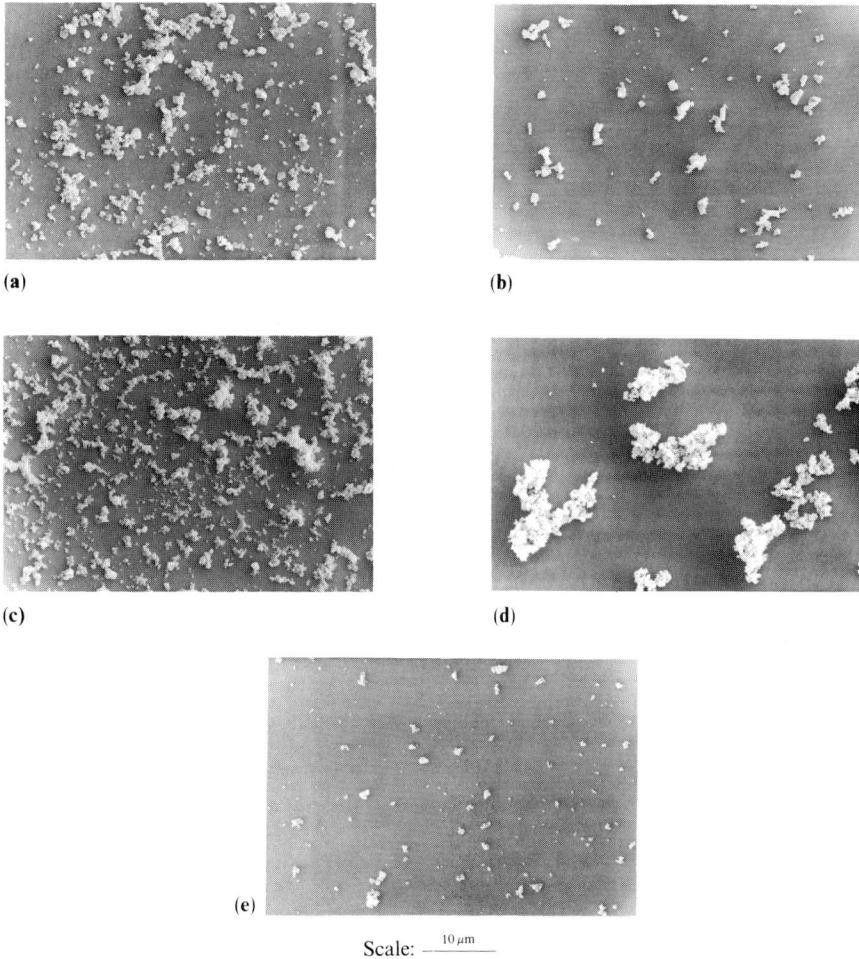

(a)

(b)

(c)

(d)

(e)

Scale: $\overline{\quad\quad}$ 10 μm

Figure 11.2   Electron micrographs of some common fillers (cont'd): (a) fine calcined kaolin; (b) aluminium trihydrate; (c) fine alumino-silicate; (d) synthetic organic pigment; (e) titanium dioxide.

mately spherical (e.g. ground calcium carbonates, titanium dioxide), or platey, either disc-shaped (e.g. kaolin, talc) or, occasionally, rod-like (e.g. aragonite).

Table 11.1 compares three model fillers with spherical, disc-shaped and rod-shaped particles of equivalent e.s.d. (and specific gravity) and gives the estimated [4] actual particle dimensions, total surface areas and particle numbers per unit mass of the disc and rod-shaped particles, compared with those for the spherical particles.

The maximum dimensions of platey disc- and rod-shaped particles are clearly significantly greater than the e.s.d.. The surface areas per unit mass are also greater than for spherical particles of equivalent e.s.d. and specific gravity, but particle numbers per unit mass are only greater for the rod-shaped particles. For disc-shaped particles, particle numbers per unit mass are somewhat less than for spherical particles of equivalent e.s.d.. It should be noted here that particle numbers per unit mass are very sensitive to particle size and this is discussed further below.

For particles that have an aggregated structure, the particle shape can usually be considered spherical, but there is a problem in defining the specific gravity. For example, an aggregated clay or calcium carbonate particle containing a void or pore volume of say $1\,cm^3\,g^{-1}$ has an actual size approximately twice that calculated assuming that the particle is solid. Surface area is dominated by the internal structure of the aggregate and is normally independent of the aggregate size. The number of aggregates per unit mass is less than that estimated from e.s.d., and in the example given would be about half the expected value.

Having accepted that the quoted particle size must be interpreted with care when considering non-spherical and aggregated particles, it is important to appreciate the general relationships between size, surface area and particle numbers per unit mass. Table 11.1 compares model fillers with particles of different shape, but of the same size. Table 11.2 shows the effect of a change in size on surface area and particle numbers per unit mass for spherical particles of the same specific gravity.

Table 11.1 Relationship between actual diameter, surface area per unit mass and particle numbers per unit mass for spherical particles of the same e.s.d. and specific gravity

	Theoretical multiplication factors for anisometric particles $(r \gg 1)$		
	Spherical	Disc-shaped $r = 20$	Rod-shaped $r = 20$
Actual diameter (disc diameter or rod length)	d	$2.96\,d$	$3.69\,d$
Surface area per unit mass	S	$2.48\,S$	$3.70\,S$
Particle numbers per unit mass	N	$0.51\,N$	$5.32\,N$

r is the aspect ratio. For a disc it is the ratio of disc diameter to disk thickness and for a rod it is the ratio of rod length to rod diameter.

Table 11.2 Relationship between actual diameter (d), surface area per unit mass (S) and particle numbers per unit mass (N) for spherical particles of the same specific gravity

Actual diameter	d	$0.5\,d$	$0.25\,d$	$0.1\,d$
Surface area per unit mass	S	$2\,S$	$4\,S$	$10\,S$
Particle numbers per unit mass	N	$8\,N$	$64\,N$	$10^3\,N$

Table 11.3 Sedimentation particle size and specific surface area data for typical fillers

Filler	Electron micrograph	Particle size/wt%				Surface area $m^2 g^{-1}$ (N_2, BET)
		$+10\,\mu m$	$-2\,\mu m$	$-1\,\mu m$	$-0.25\,\mu m$	
Standard kaolin	1	10	50	35	8	9
Fine kaolin	–	0	80	60	15	12
Standard chalk	2	5	45	25	1	3
Ground limestone	3	1	60	35	10	7
Scalenohedral precipitated calcium carbonate	4	0	80	50	10	7
Talc	5	30	17	5	0	6
Fine calcined kaolin	6	1	90	70	5	14
Aluminium trihydrate	7	0	90	60	8	7
Fine alumino silicate	8	8	70	55	25	65
Coarse alumino silicate	–	40	40	35	15	55
Synthetic organic pigment	9	10	5	0	0	22
Titanium dioxide (anatase)	10	0	96	94	8	23
Silica	–	6	40	30	18	150

For a given shape and specific gravity, surface area is proportional to $1/(\text{size})$ whereas particle numbers per unit mass are proportional to $1/(\text{size})^3$. Assuming that non-spherical particles may be described as discs, for particles of the same disc diameter and specific gravity, surface area and numbers per unit mass are proportional to aspect ratio r.

From Tables 11.1 and 11.2 it can be seen that the effects of particle size and shape on specific surface area are of the same order, and for a given shape a reasonably large difference in mean particle size is required to give the same effect as the difference, say, between platey and blocky particles. For particle numbers per unit mass, however, the effect of particle size clearly dominates. Thus, phenomena influenced by surface area will be influenced by the shape and structure of filler particles as well as by particle size, whereas phenomena dependent on particle numbers will depend far more on size than on shape.

Table 11.3 gives some sedimentation particle size and specific surface area data for typical fillers including those shown in the electron micrographs. The particle size distributions typically lie in the range ($< 10\,\text{wt}\% > 10\,\mu m$) – ($< 10\,\text{wt}\% < 0.1\,\mu m$) e.s.d. with the exception of titanium dioxide which has a relatively fine particle size distribution.

11.2.2 Surface chemistry

All paper fillers disperse readily in water and this implies that the particle surfaces are either naturally hydrophilic or are coated with a surfactant. Of the natural mineral fillers, talc is the least hydrophilic and in fact the slightly hydrophobic nature of talc has led to its common use as a pitch adsorbent in papermaking systems.

There has been considerable discussion of the importance of electrophoretic mobility and zeta potential to the retention of fillers and fines [5]. However, studies of flocculation (i.e. particles held together primarily as a result of bridging by long chain polyelectrolytes often of the same charge as the particles) and coagulation (i.e. particles held together as a result of mutual attraction in the absence of electrostatic repulsion) demonstrate that zeta potential gives, at best, only an approximate indication of particle behaviour [6]. This assumes that problems in measurement and the interpretation of electrophoretic mobility measurements may be overcome [7].

It is universally accepted that a zeta potential of at least 20 mV is sufficient to prevent coagulation of similarly charged particles [8], and that a zeta potential of greater than about 40 mV is sufficient to prevent flocculation by non-ionic polymers or by polymers of similar charge to the particles [9]. Obviously, increasing the particle charge also increases the amount of oppositely charged polymer required to induce flocculation or coagulation.

Zeta potential data[2] are given in Table 11.4 for fibre and for several fillers. The data are mainly taken from the literature and refer to materials in clean systems (typically distilled water with low levels of monovalent ions) without the addition of charge-determining ions or polyelectrolytes. Even under these conditions, coagulation of filler particles with either fibre and fibre fines or with other particles may be expected to occur for many fillers. In practical papermaking systems, charge-determining ions and polyelectrolytes are present and the zeta potential of the filler particles can be considerably different from that measured in a clean system [13–15]. Table 11.5 gives some examples of zeta potential measurements for fillers in environments closer to those expected in practical systems. Measurements of zeta potential of filler particles in the presence of fibres indicate that the adsorption of ions, polyelectrolytes or colloidal material associated with the fibre leads to a zeta potential value for the filler remarkably similar to that of the fibre. Clearly the electrophoretic properties of the 'pure' fillers are of limited importance and the adsorption properties of the filler particles may well be of greater significance.

The only practical circumstances under which coagulation or flocculation of the filler with the fibre and fibre fines may prove difficult is when the filler particles carry a high negative surface charge. This may arise, for example, from the need to stabilise the filler as a high solids content slurry for transportation, or from the return of coating pigments as filler to the wet end of

---

[2] Several authors quote electrophoretic mobility rather than zeta potential. This correctly avoids the use of theoretical expressions to convert the experimental observations of mobility into a zeta potential. When zeta potential is quoted in the literature the Smoluchowski equation is generally assumed to apply, and for comparative purposes this equation has been used to convert mobility measurements to zeta potential in Tables 11.4 and 11.5.

Table 11.4   Zeta potentials of fibre and fillers

Material	Measurement conditions	Reference	pH	Zeta potential (mV)
Fibre	Distilled water $+ H_2SO_4$	13, 19	4	-15
	or NaOH to adjust pH		6	-25
			8	-30
Filler				
Kaolin	20 ppm, 10^{-3} mol dm^{-3} NaCl	10, 11	5	-20
			7	-30
Titanium dioxide	Distilled water $+ H_2SO_4$ or	13	4	-9
anatase	NaOH to adjust pH		6	-28
	Deionised water	12	6.4	-48
rutile	Distilled water $+ H_2SO_4$ or	13	4	$+25$
	NaOH to adjust pH		6	$+4$
Hydrated aluminium	Distilled water $+ H_2SO_4$ or	13	6	$+17$
anatase	NaOH to adjust pH		8	$+9$
Calcium carbonate				
chalk	200 ppm, distilled water	15	–	-20
marble	200 ppm, distilled water	15	–	0
PCC	1000 ppm, distilled water $+ 10$ ppm, Ca^{2+}	16	–	$+12$
Silica	Distilled water	17	5	-40
Talc	Distilled water $+ H_2SO_4$ or	18,23	4	-10
	NaOH to adjust pH		6	-20

the paper machines in mills producing coated paper. Zeta potentials as high as -40 mV have been recorded under these circumstances [18].

It is obviously important that polymeric flocculants or retention aids are adsorbed by the filler particles. Interestingly, most fillers will adsorb both cationic and anionic polymers under normal papermaking conditions [24]. Non-ionic polymers are also adsorbed. With the exception of pure calcite (i.e. not contaminated with natural organic acids or silicates) and some organic pigments, greater amounts of cationic polymers than of anionic polymers are normally adsorbed. The amounts adsorbed increase with increasing surface area [25], and, where insufficient surface sites exist for the polymer, adsorption is also increased by ions or polymers of opposite charge to that of the polymer [26].

Surface area is particularly important if a polymer of opposite charge to that of the filler is required to reduce an excessively high surface charge. Obviously the amount of polymer required will increase with increasing surface area.

The surface area is also relevant to the adsorption of other paper chemicals, in particular sizing materials. The adsorption of large amounts of paper chemicals by the filler may be detrimental to paper properties and this is discussed further below.

Table 11.5 Effect of charge-determining ions and polyelectrolytes on zeta potential of fillers (filler concentration $\sim 200\,ppm$)

Filler	Solution or pigment treatment	Reference	Zeta potential (mV)
Kaolin	In presence of pulp fines $+ H_2SO_4$ to pH 6	–	-15
	0.5 wt% PEI (on filler), distilled water $+ H_2SO_4$ to pH 6	19	$+22$
	$0.2 \times 10^{-6}\,mol\,dm^{-3}$ sodium lignosulphonate $+ H_2SO_4$ to pH 6	10	-35
	$0.6 \times 10^{-4}\,mol\,dm^{-3}$ $Al_2(SO_4)_3$ + NaOH to pH 6	11	$+15$
Titanium dioxide anatase	In presence of pulp fines $+ H_2SO_4$ to pH 6	13	-15
	In presence of pulp fines + $7 \times 10^{-6}\,M\,Al_2(SO_4)_3$ to pH 6	22	$+5$
	1 wt% PEI (on filler), distilled water $+ H_2SO_4$ to pH 6	20	0
Calcium carbonate marble	$NaTPP\ 10^{-4}\,mol\,dm^{-3}$	21	-30
limestone	$NaTPP\ 10^{-4}\,mol\,dm^{-3}$		-25
PCC	In presence of pulp fines + 30 ppm Ca^{2+}	16	-15
Hydrated aluminium oxide	In presence of pulp fines $+ H_2SO_4$ to pH 6	13	-13

11.3 Retention

11.3.1 *Mechanical entrapment*

In simple terms, filler is retained mainly by mechanical entrapment if the first pass filler retention[3] is below 30%. This applies, for example, to high speed machines for which shear rates or the content of ionic species in the system are such that retention aids are ineffective, and machines for which constraints on paper quality, such as formation, preclude the use of highly efficient retention aid systems. Studies of the wet-end systems of high speed magazine and newsprint machines have confirmed that mechanical entrapment can be a dominant mechanism [27].

For mechanical entrapment to dominate, electrokinetic interactions must be insignificant, and filler particles are simply trapped in the pore structure of the paper web as it is formed on the wire. As expected, retention increases with

---

[3] First pass retention (%) $= \dfrac{F_H - F_{ww}}{F_H} \times 100$

where F_H = concentration of filler in flow box
F_{ww} = concentration of filler in white water

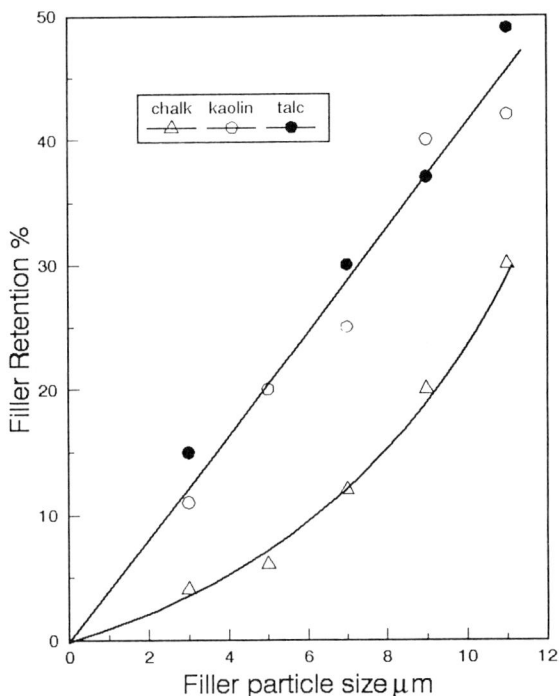

Figure 11.3   Filler retention in a 60 gsm handsheet. Bleached softwood sulphite pulp (300 CSF) + 30% w/w filler + 0.1% w/w polyacrylate dispersing agent.

the particle size of the filler, and since the actual particle dimensions are more relevant, platey fillers are retained to a greater extent than blocky fillers of the same e.s.d. (Figure 11.3). Similarly, aggregated pigments can be retained to a greater extent than expected from the e.s.d. Retention also increases as the freeness of the pulp decreases (Figure 11.4). From a theoretical viewpoint the retention of mono-sized latex or similar particles could give useful information on the pore structure in paper webs.

The preferential retention of coarse particles does not necessarily lead to differences between the net particle size distribution of the filler in the sheet, compared with that added to the machine system. The proportion of fine filler particles in the white water system and flow box does, however, increase, and variations in both the filler particle size distribution and the filler mass distribution in the z-direction of the sheet can occur.

11.3.2   *Electrokinetic effects*

Electrokinetic effects begin to dominate as the first pass filler retention increases above 30%. Coagulation and flocculation of filler with fibres and

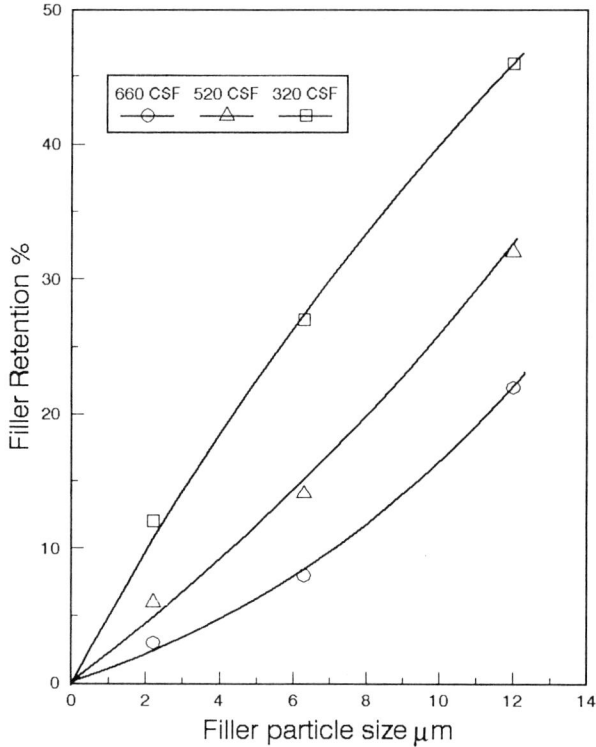

Figure 11.4 Effect of pulp freeness on filler (kaolin) retention. Forming conditions as for Figure 11.3.

fibre fines become the major retention mechanism. This is the case for many machines using modern retention aids.

Recent work with model systems [28–31] has shown that the available surface area of the fibre and the surface area of the filler (or particle size) are important. Under shear conditions an equilibrium exists between deposition of particles on the fibre surfaces and detachment of particles from the surface. Deposition is driven by attractive electrokinetic forces created by the retention aids whereas detachment occurs through hydrodynamic shear. Retention can be described by Langmuir type adsorption models with a maximum adsorption capacity for filler particles on the fibre surfaces. In one system [29] black spruce pulp was shown to have a capacity of about 0.2 g/g of a titanium dioxide filler at low levels of beating increasing to almost 0.6 g/g at high levels of beating. Presumably for fillers such as fine kaolins with high surface area, the adsorption capacity could be as low as 0.2 g/g (i.e. 20% w/w) in many real papermaking systems. This is illustrated by the optical micrographs in Figure 11.5 which show that at 30% w/w of a

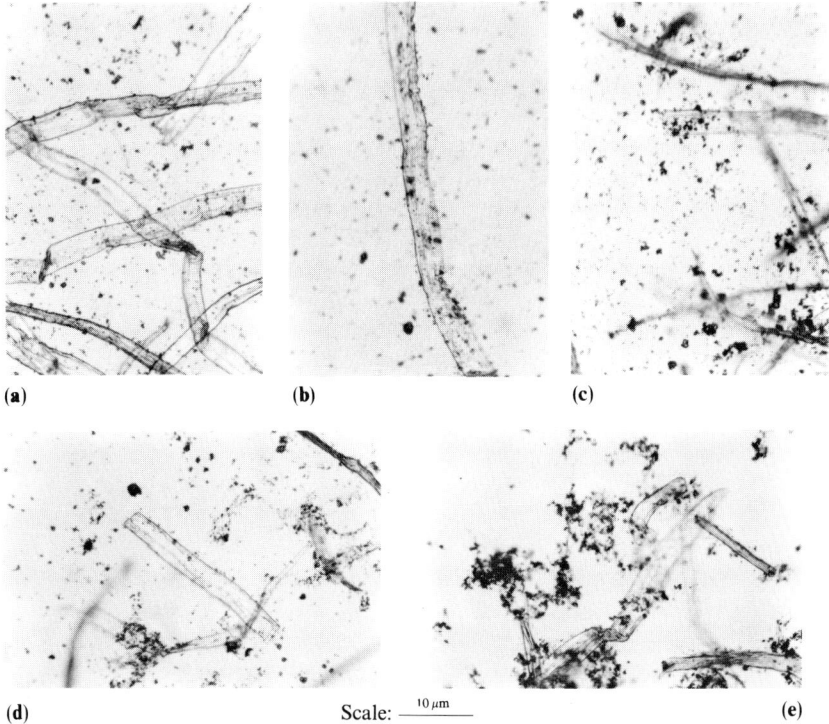

(a)   (b)   (c)

(d)   Scale: ___10 μm___   (e)

Figure 11.5 Optical micrographs of pulp systems containing 30% w/w filler with and without retention aids: (a) unbeaten pulp (no retention aids); (b) unbeaten pulp ('strong' retention aid); (c) beaten pulp (300 CSF) (no retention aid); (d) beaten pulp (300 CSF) ('weak' retention aid); (e) beaten pulp (300 CSF) ('strong' retention aid).

conventional kaolin filler, a typical level in the flow box of some highly filled paper machines, there could be more filler particles present that can be attached to the fibre surfaces even in the presence of 'strong' retention aids (see Figure 11.5(b)). In such cases retention by mechanical entrapment of filler particles or filler aggregates could remain an important mechanism.

In most practical situations the retention level become a balance between the machine speed, and hence shear rates in the flow to the wire, and during drainage, the efficiency of the retention aid system and the paper formation or rather the degree of fibre flocculation acceptable.

Calculations of the shear rate at various points in a typical paper machine are summarised in Tables 11.6 and 11.7. At a machine speed of $1000 \, \text{m min}^{-1}$ shear rates of 10^3–$10^4 \, \text{sec}^{-1}$ are not uncommon [32].

The Dynamic Drainage Jar and similar techniques have been extensively used to demonstrate the effects of retention aid systems [33]. Briefly, the required furnish is stirred under controlled conditions and, after a suitable

Table 11.6 Shear rates (sec^{-1}) in papermaking systems measured at two different machine speeds ($m\,min^{-1}$) [32]

Component	Shear rate (sec^{-1})	
	610 m min^{-1}	1000 m min^{-1}
Pressure screen	2×10^3	2×10^3
Fan pump	3×10^3	4×10^3
Slice	4×10^2	7×10^2

For comparison, the Dynamic Drainage Jar operating at a speed of 500 rpm has a shear rate of $2 \times 10^3 \, sec^{-1}$.

Table 11.7 Maximum shear stress (Pa) on fibre walls measured at two different machine speeds ($m\,min^{-1}$) [32]

Component	Shear stress (Pa)	
	610 m min^{-1}	1000 m min^{-1}
Table rolls	7×10^2	2×10^3
Foils	4×10^2	1×10^3
Pressure screen	1×10^4	1×10^4
Fan pump	2×10^4	2×10^4
Slice	80	2×10^2

For comparison, the Dynamic Drainage Jar operating at a speed of 500 rpm has a shear stress of 3×10^2 Pa.

period, a portion of the supernatant is drawn off through a screen. Normally no web formation on the screen is allowed, so that the proportion of fillers and fine flocs (depending on the screen aperture size) in the supernatant allows estimation of the proportion of filler attached to coarser fibre particles. This is taken as the filler retention. Figure 11.6 gives some typical results. The range of stirring speeds normally covers the shear rates expected in machines of speed 750–1000 m min^{-1} (see Tables 11.6 and 11.7). Dynamic Drainage Jar experiments are at best semi-quantitative, but give a valuable indication of the effects of various additives in real papermaking systems.

From a fundamental point of view the complexity of the papermaking system has led to much discussion of retention mechanisms, although a general understanding is beginning to emerge based on the general theories of colloidal chemistry and on model experiments.

At the shear rates normally encountered in paper-machines, either flocculation through bridging polymeric retention aids or strong coagulation through 'patch charge' effects appear essential [34]. More sophisticated retention aid systems such as the micro-particulate and other two component systems are even more effective [35]. Presumably the essential effect is that, for good retention, small particles colliding with or approaching larger particles, or indeed other small particles, must have a high probability of attachment

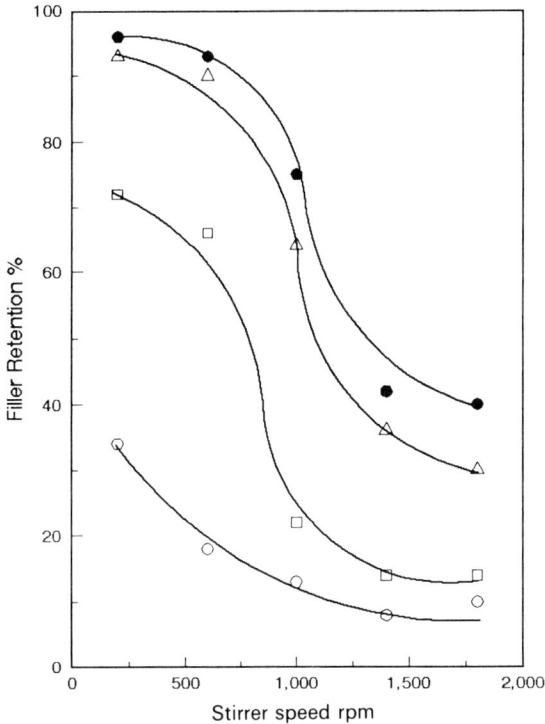

Figure 11.6   Filler retention in typical Drainage Jar experiment. Standard kaolin filler (30% w/w), bleached softwood sulphite pulp (300 CSF). (○ No retention aid; □ 0.02% anionic polyacrylamide (MW ~ 10⁶, charge density 0.1); △ 0.02% cationic polyacrylamide (MW ~ 10⁶, charge density 0.2); ● dual polymer: 0.1% cationic polyacrylamide (MW 10⁵, high charge density) + 0.02% anionic polyacrylamide).

either through electrostatic attraction or attachment to extended polymer chains. Whether the attachment should be reversible or irreversible remains a point of discussion. However, as the shear during drainage on the machine wire is not as great as that experienced at points in the flow to the wire, a degree of reversibility may be advantageous.

Theory predicts that flocculation or coagulation in the papermaking system is primarily orthokinetic, so that the number of collisions (J) for a particular particle of radius a_i with other particles of radius a_j per unit time and volume is given by [36]:

$$J = \tfrac{4}{3} G(a_i + a_j)^3 N_j$$

where N_j is the number of particles of radius a_j and G is the shear rate. On a size basis, a fibre will collide far more frequently with other fibres, fibre fines and with filler particles than a filler particle will collide with other filler particles. If a polymer is added it may be considered as consisting of very small particles which will collide far more frequently with fibre and fibre fines than with filler particles [37].

Therefore, assuming that strong interactions occur, it is expected that any polymer added to the system will be taken up primarily by the fibres–preferential adsorption by the filler is unlikely. Also filler–filler aggregates are unlikely to form, instead filler–fibre and filler–fibre fines aggregates will form. Furthermore, the order of addition of polymer and filler to the fibres over a short time scale ($\sim$ 1 sec) should be immaterial.

After formation, the aggregates must survive further shear in the paper machine or reform after points of high shear. Again, theory predicts that disruption by shear is size related: larger aggregates will be preferentially disrupted and very small filler particles should be difficult to remove from fibre surfaces [39].

There are examples of studies that support this general picture. Measurements of the rates of adsorption of cationic polyacrylamides onto kaolinite [40] and cellulose fibres, under conditions not too far from those of the papermaking system, have been made [41]. The flocculation of the kaolinite was relatively slow with an incubation period of several seconds attributed to the slow adsorption of the polymer. The flocculation of the cellulose fibres was, however, far more rapid, and was essentially complete in less than one second. These experiments support the predictions for aggregate formation. Measurements of floc disruption have also been made and, more interestingly, measurements of the shear required to remove filler particles from cellophane surfaces (Table 11.8). The dependence on particle size is clearly shown.

The reduction in shear stress required to remove coarse filler particles from fibre surfaces implies that retention of coarse filler by electrokinetic mechanism could be poor. Figure 11.7 shows that for typical papermaking systems this is not the case, fine fillers have lower retention even under optimum laboratory conditions.

This is not inconsistent with theoretical ideas and is a consequence of the amount of fine fibre fragments and fibrillation present. The filler particles are rarely attached to the actual fibre surface in the wet state, but are

Table 11.8 Shear stress (Pa) required to detach filler particles of three different diameters (μm) from cellulose [38]

Retention aid	Shear stress (Pa)		
	0.4 μm	1 μm	1.4 μm
None (pH 7)	28	6	4
None (pH 3)	170	39	22
Cationic starch (pH 7)	330	75	44
Polyethylenimine (pH 10)	850	19	110
Cationic polyacrylamide (pH 7)	2200	510	290

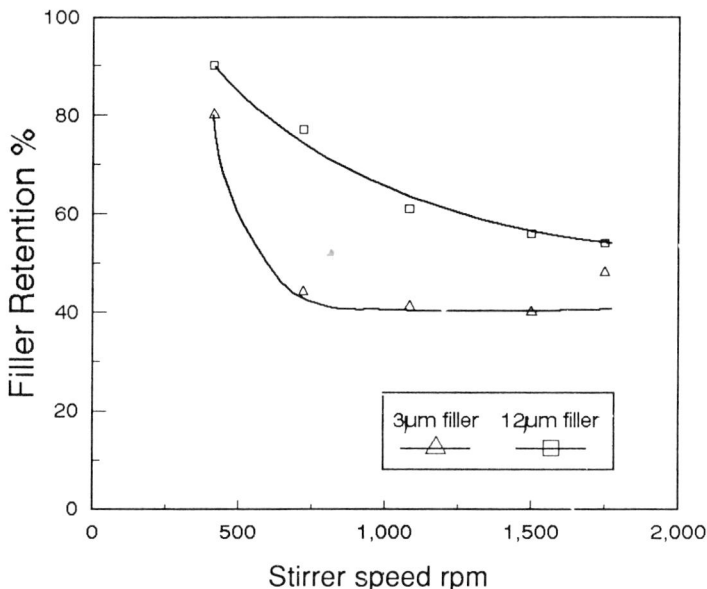

Figure 11.7   Drainage Jar comparisons between 3 μm and 12 μm kaolin fillers. Bleached softwood sulphite pulp (300 CSF), 30% w/w filler, 0.4% w/w polyethylenimine.

attached to fine fragments and fibrils which are either part of the fibres or are themselves attached to the fibres by the same electrostatic forces that fix the filler particles (see Figure 11.5). In the presence of a retention aid, the material passing through the machine wire (which has apertures of the order of 100 μm) is most likely filler-fibre fines aggregates rather than free filler or fibre fines particles [36, 42]. Therefore the retention controlling step is probably the disruption of relatively large filler-fibre fines aggregates rather than the removal of individual filler particles from the main fibres. The strength of attachment of these aggregates to the main fibres may be reduced more significantly by the presence of finer filler particles due simply either to the presence of a greater number of particles preventing interaction between the fibres fines and the fibre (analogous to the Langmuir type theories), or to a higher filler surface area adsorbing retention aid.

There has been some discussion as to the preferred order of addition of retention aids. It has been suggested that the filler should be pre-treated with a highly-charged cationic polymer to disperse, cationically, the filler particles and to promote strong coagulation with negative fibre particles [19]. However, in practice this does not give particularly good retention. Pre-treatment of the filler with polymeric flocculants to form filler–filler flocs greatly influences the effect of the filler on sheet properties as discussed below, but again does not significantly improve retention. Pre-treatment of the fibres has also been suggested and some benefits are observed [43]. In general, these practical observations are not inconsistent with theory.

11.4   The effect of filler on paper properties

Changes in paper properties caused by the addition of a filler are due partly to the nature of the filler itself and partly to the disruption of the fibre network as a result of the inclusion of the filler particles.

Examples of the direct effect of the filler include adsorption of sizing chemicals by the filler, which may reduce the degree of paper sizing; light absorption or scattering by filler particles, which may influence the optical properties of the paper; and absorption of ink by the internal pore structure of filler particles, which may reduce ink penetration into the paper. Examples of the effect of disruption of the fibre network include loss of paper strength, increase in bulk per unit fibre mass and an increase in the light scattering coefficient of the fibre network.

The effect of fillers on paper properties must be considered therefore as two separate subjects:

(i)   disruption of the fibre network or filler–fibre interaction
(ii) the intrinisic properties of the filler.

11.4.1   *Interaction between filler and fibre*

The nature of filler–fibre interactions can be inferred from microscopic examination of the furnish and the finished paper, and from an analysis of paper strength, bulk (volume per unit mass) and light scattering coefficient.

The optical micrographs of papermaking furnishes shown in Figure 11.5 show that, for a paper with reasonable filler retention (say $> 50\%$ first pass filler retention) and no significant aggregation of the filler prior to addition to the fibre suspension, the filler is retained almost entirely as loose aggregates of fibrils, fibre fines and filler particles. In the finished paper these aggregates appear either as free aggregates in interstices between principal fibres, or are trapped between fibres (Figures 11.8 and 11.9). Experiments with fillers that may be removed from the finished sheet by dissolution without further change in the fibre network clearly demonstrate the fibrillar content of these aggregates [44] (Figure 11.10).

Paper strength, particularly burst and tensile strength, is reduced by the filler, and the paper bulk per unit mass of fibre is increased. This is attributed to a reduction in bonded area as a result of the trapping of filler, fibril and fibre fines flocs between fibre areas which would normally have bonded.

The influence of fibre–filler–fibre bridging as a contribution to paper strength can probably be ignored. Experiments with cationic polymers

(a)

Scale: ——— 50μm ———

(b)

Scale: ——— 10μm ———

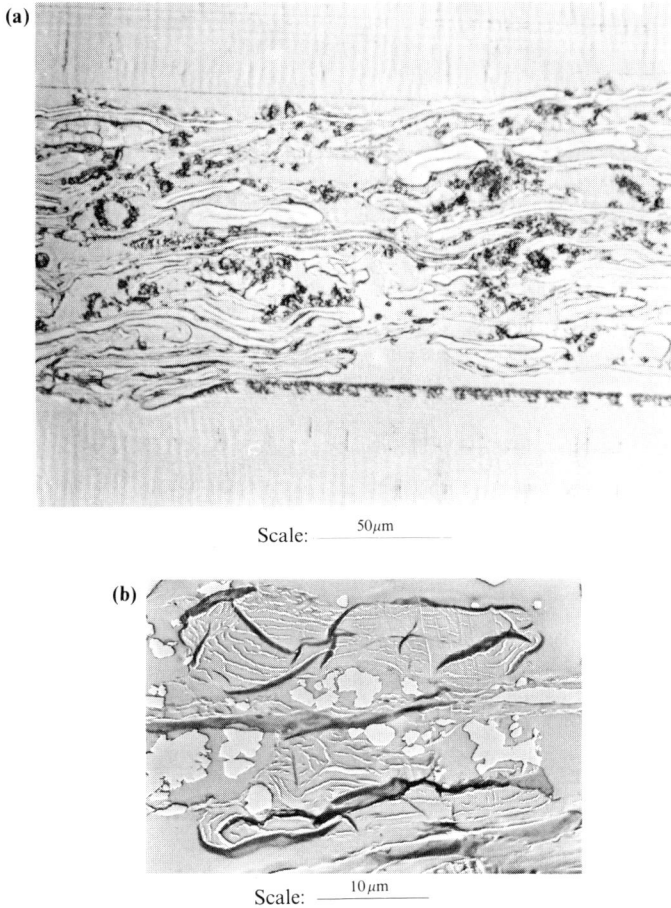

Figure 11.8 Optical (a) and transmission electron (b) micrographs of cross-sections of chalk-filled paper. Note: in (b) chalk particles dissolve during sample preparation and appear as holes.

[19,45], talcs, ground starch particles and various organic fillers [44,46] confirm this. In most cases, provided due allowance is made for any differences in specific gravity and comparisons are made between fillers of the same particle size and shape, the effects of fillers on paper strength are independent of surface chemistry. Any exceptions are normally due to irreversible aggregation of the filler prior to addition to the fibre and this modifies the apparent particle size of the filler. An interesting study, that of the effect of montmorillonite as a filler [47], showed that a very fine (< 0.1 μm) extremely platey and flexible filler with a hydrophilic surface has little deleterious effect on paper strength. The light scattering coefficient of the paper is, however, not

(a)

Scale: ___50 μm___

(b)

Scale: ___10 μm___

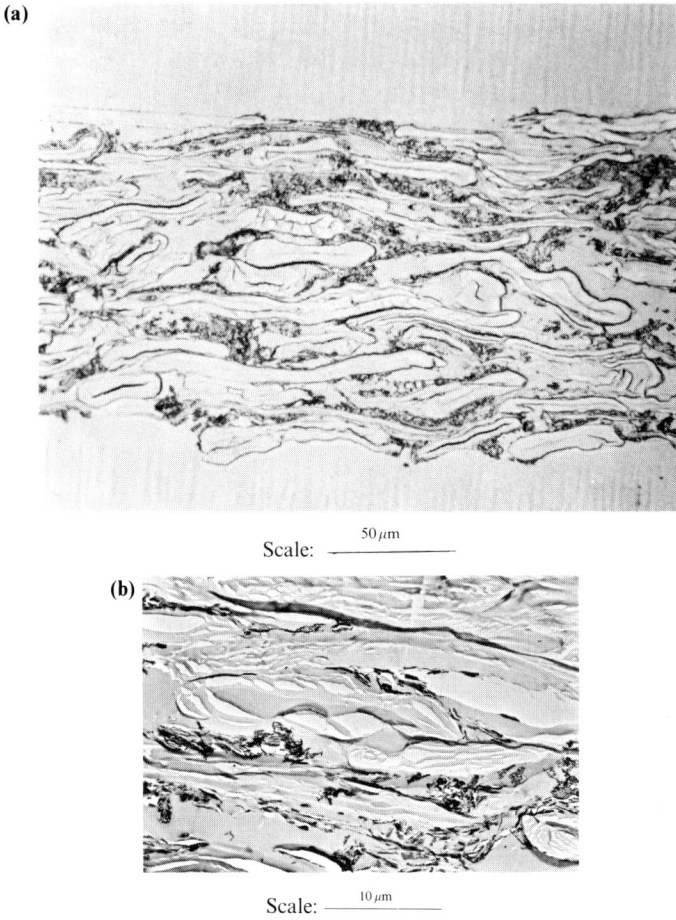

Figure 11.9 Optical (a) and transmission (b) electron micrographs of cross-sections of clay-filled paper.

increased, indicating that the filler is in close contact with the fibre and that the bonded area of the fibre remains the same. The filler particles conform to the surface of the fibre, even during the drying stage when the fibres shrink, and form strong filler-fibre bonds. The rigid nature of conventional filler particles prevents such conformation [48] and hence filler-fibre bonds are relatively ineffectual. The light scattering coefficients of papers containing conventional fillers indicate that at least 60% of the filler surface area is free from fibre contact and confirm that the bonded area of the fibre is reduced [44, 49]. Fibre–filler–fibre bonding need only be considered if the light scattering coefficient of the paper is unusually low.

It is proposed that the loose flocs of filler, fibrils and fibre fines are flexible,

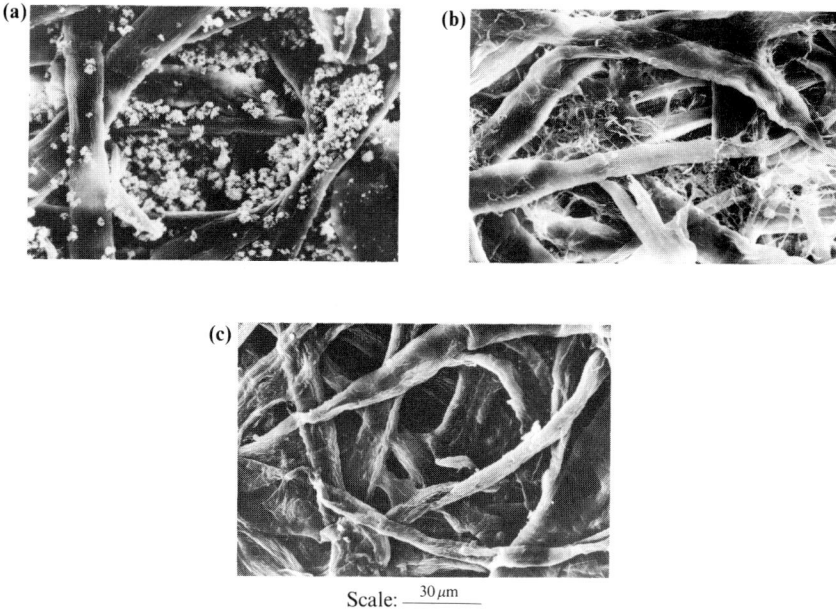

Scale: $\underline{30\,\mu m}$

Figure 11.10   Effect of filler removal on fibre network: (a) filled sheet; (b) filler removed by dissolution in HCl/EtOH; (c) unfilled sheet. Note: HCl/EtOH showed no change in structure.

but after initial compression reach a fixed packing density that is dependent on the shape, and possibly size, of the filler particles, and which remains constant throughout subsequent stages of paper formation. The flocs, however, remain flexible, and as the paper consolidates during pressing and drying, are deformed by the applied pressure (and perhaps surface tension forces drawing fibres together) and spread out over the fibre surfaces. Finally, at a certain stage for some reason, no further deformation is possible and the structure is fixed. The remaining water is removed from the floc structure by thermal drying. Fibre bonding is prevented over the area occupied by the floc.

Assuming that, with the addition of a filler and in the absence of any changes in papermaking conditions, the bulk (or volume per unit fibre mass) of the fibre network is increased and paper strength is decreased only by separation of bonded areas, then a relationship must exist between bulk and strength. The simplest relationship would be:

$$\frac{\Delta V}{\Delta A} \propto t,$$

where ΔV is the increase in paper volume per unit mass of fibre, ΔA is the loss of paper strength (tensile or burst) and t is the distance of separation between areas de-bonded by the flocs. This distance is in effect the average thickness of the floc in the finished sheet.

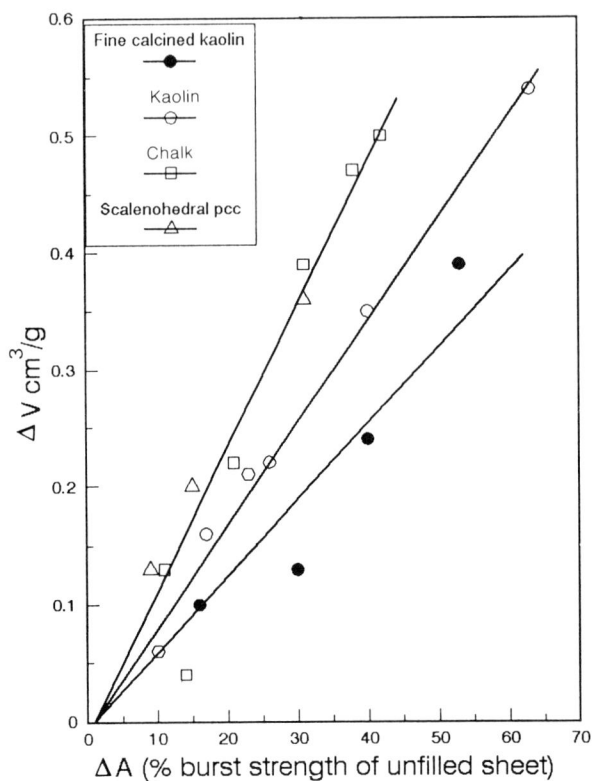

Figure 11.11 Plots of ΔV (increase in sheet bulk, g^{-1} fibre) against ΔA (decrease in sheet strength, g^{-1} fibre).

Table 11.9 Estimate of spacing (t) of de-bonded fibre areas in filled handsheets

Filler	Micrograph figure	$t\,(\mu m)$
Standard kaolin	1	3.0
Fine kaolin	–	1.5
Standard chalk	2	4.0
Scalenohedral precipitated calcium carbonate	4	4.0
Fine calcined clay	6	2.3
Aluminium trihydrate	7	2.5
Synthetic organic pigment	9	6.0

Figure 11.11 shows some typical plots of ΔV against ΔA for various fillers in softwood sulphite handsheets prepared under standard conditions (Tappi 205). The pulp was beaten to 300 CSF. The data points represent different loading levels and the plots are reasonably linear confirming the independence of t from filler loading. For this particular pulp system and the handsheet preparation conditions used, the bonded area without filler is about $0.5\,m^2\,g^{-1}$. This allows an estimation of values of t, and some results are given in Table 11.9.

An interesting comparison may be made between specially prepared narrow particle size distribution fillers. Figure 11.12 shows plots of t against e.s.d. for kaolin and calcium carbonate fillers. Allowing for particle shape, t is approximately equivalent to the actual dimensions of the particles. This implies that, under the conditions of formation, the compression of the flocs is stopped by physical jamming of particles between fibres. For the same fillers,

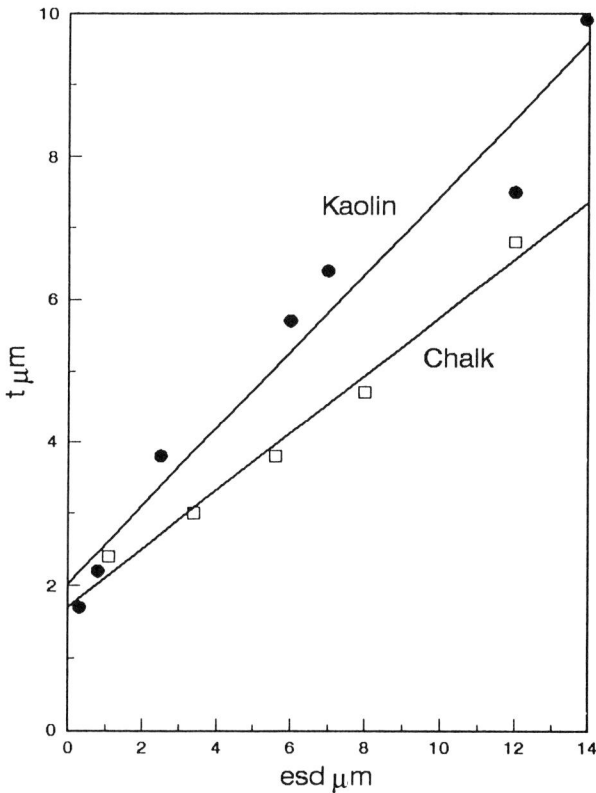

Figure 11.12   Plot of separation (t) against mean e.s.d. for narrow particle size distribution fillers.

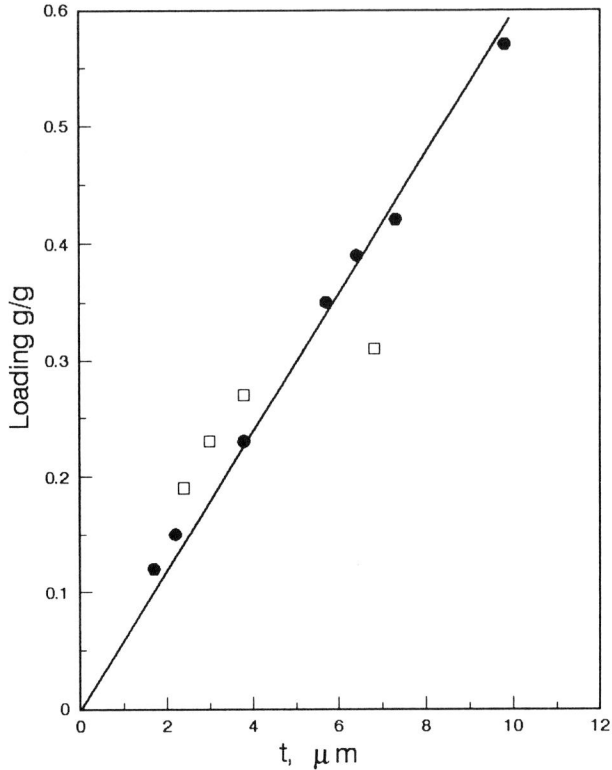

Figure 11.13 Plot of filler loading required to reduce sheet burst strength (per unit fibre weight) by 30% against t.

Figure 11.13 shows plots of the filler loading required to give a fixed loss of paper burst strength against t. These plots are reasonably linear and indicate that the packing densities (volume occupied per unit mass of filler) of these fillers are approximately independent of size and shape. As is generally observed in commercial practice, the effect of kaolin and calcium carbonate filler on paper strength increases as the particle size decreases.

Paper strength is often related to the specific surface area of the filler. As discussed earlier, surface area is proportional to 1/particle size, depending on particle shape. Therefore in any relationship, particle size may be replaced by 1/specific surface area. However, the particle shape must be accounted for, for example platey fillers have a substantially higher surface area compared with blocky fillers of the same particle size. This, together with the relationship between particle size and paper strength loss, explains the apparently highly detrimental effect of fillers with blocky particles such as calcium carbonate, when compared per unit specific area with platey fillers such as kaolin or talc.

Returning to Table 11.9, a comparison of the values of t with the electron micrographs of filler particles shown earlier confirms that, for most fillers, the physical dimensions of the filler particles determine the separation of fibres in the de-bonded areas. The combination of the separation, t, and the ultimate packing density of particles, determines the loss of paper strength and increase in bulk of the fibre network. For most kaolin and carbonate fillers, the packing density in the flocs is obviously very similar. For some fillers, however, the packing density can be quite different. Aggregated fillers such as calcined kaolin and scalenohedral precipitated calcium carbonate have relatively large effects on strength and bulk compared with non-aggregated kaolins and calcium carbonates of a suitable particle size to give the same fibre separation (Figure 11.14). The packing density is relatively low due partly to the internal porosity of the aggregated structures, and partly to the poor packing of the aggregates. Conversely, conventional fillers aggregated with a polymer flocculant prior to addition to the fibre, have a relatively small effect on strength and bulk (Figure 11.15) and a higher packing density. Filler aggregates formed by polymer flocculants [50] are large ($\sim 50\,\mu m$) and compressible. The typical fibre spacing generated is about $10\,\mu m$. Presumably, compared with the normal flocs of filler particles with fibrils and fibre fines, in the filler-only flocs the filler particles may pack more effectively. The lubricating effect of the flocculating polymer in releasing stress in the sheet is also seen as important [52].

The effect of a filler on paper strength and bulk is always apparent in the light-scattering coefficient of the paper. Greater disruption of the fibre network leads to a higher light-scattering coefficient. This is due to the greater area of fibre and fibrils or fibre fines available to scatter light.

The difference between the effect of 'rigid' aggregation, such as that formed by intercrystal growth during precipitation or by fusion during calcination of kaolin, and 'soft' aggregation such as that formed by polymer flocculation, has important practical consequences in papermaking. Depending on the size of the aggregate formed, rigid aggregation may increase the disruption of the fibre network whereas soft aggregation invariably reduces the disruption. More importantly, rigid aggregation also preserves an open structure for the filler aggregate in the finished paper. This increases the light-scattering coefficient of the filler in the paper and is discussed further below. Rigid aggregation is preferred when a substantial increase in the light-scattering coefficient of the paper is required. The use of calcined clays, precipitated alumino-silicates and some precipitated calcium carbonates are all examples of this approach. Conversely, soft aggregation is preferred when an increase in filler loading is required and this has led to considerable research on the soft flocculation of fillers prior to addition to the fibre [51]. Essentially, any irreversible soft aggregation of the filler will lead to some increase in filler loading for a given strength loss. An intriguing, but seemingly fortuitous, consequence of soft aggregation is that for a given filler and paper strength, the

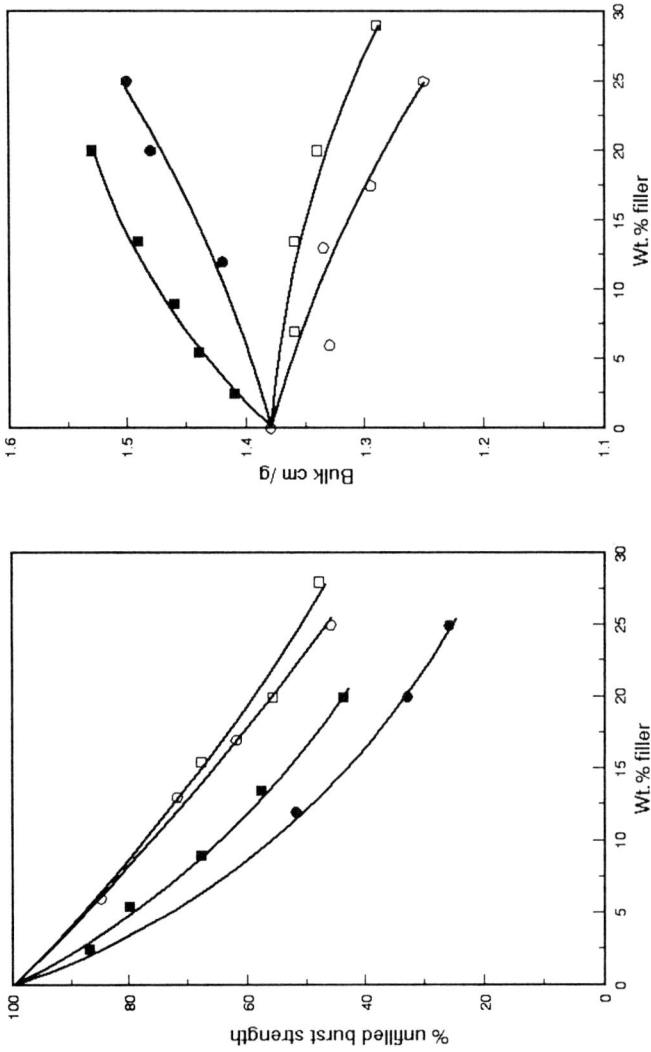

Figure 11.14 Plots of burst strength and bulk against filler content (w/w%) for aggregated calcium carbonate (scalenohedral pcc) (■) and calcined kaolin (●) compared with standard calcium carbonate (□) and kaolin (○). Bleached sulphite softwood pulp (300 CSF), standard handsheets, 0.02% w/w polyacrylamide retention aid, giving similar values of fibre separation ($t = 2.5$ and $4.0 \mu m$ for kaolin and calcium carbonate, respectively).

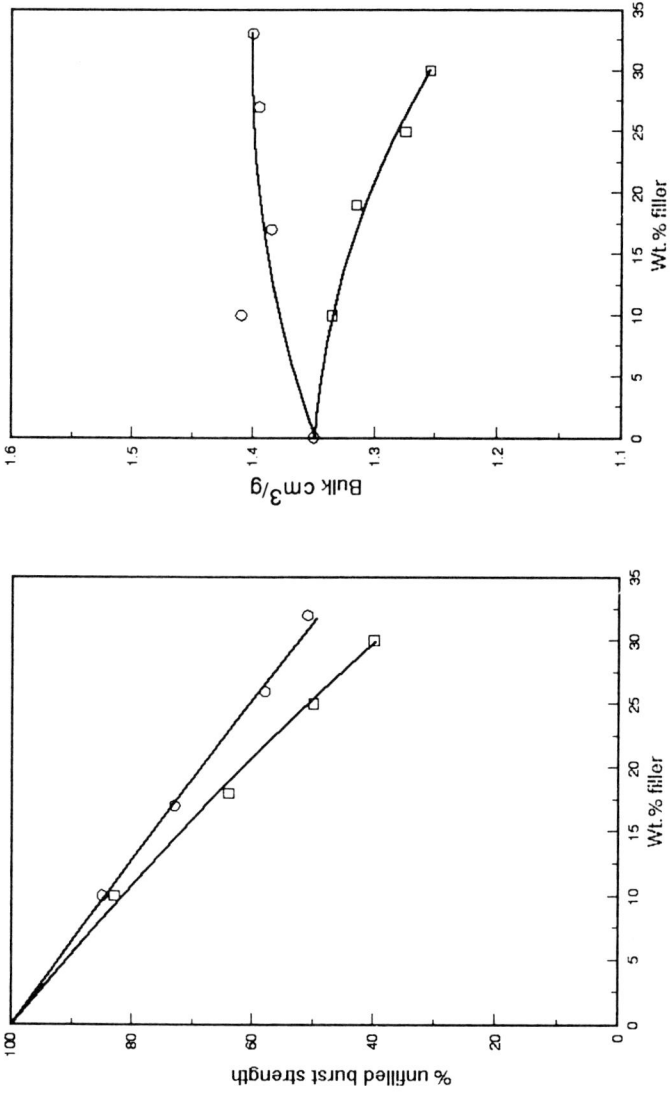

Figure 11.15 Plots of burst strength and bulk against filler content (w/w%) for polymer flocculated kaolin (□) compared with a coarse kaolin (○) giving the same fibre separation ($t = 10\ \mu m$).

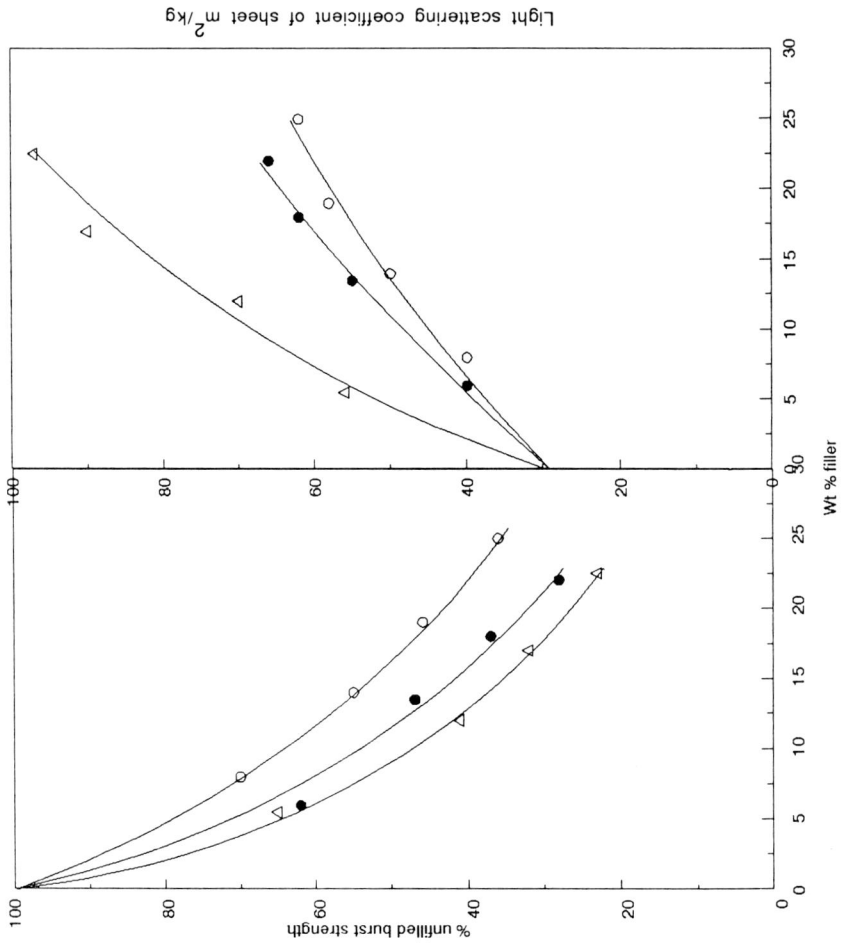

Figure 11.16 Plots of burst strength and sheet light-scattering coefficient against filler loading (w/w%) for an untreated fine kaolin (●) compared with the effects of 'hard' aggregation (calcination) (△) and 'soft' aggregation (polymer flocculation) (○).

paper light-scattering coefficient remains approximately constant as the filler floc size and hence possible filler loading is increased [44, 52]. Figure 11.16 summarises the effects of rigid (calcination) and soft aggregation on paper strength and light-scattering coefficient for a fine kaolin filler.

11.4.2   The intrinsic properties of fillers

11.4.2.1   *Surface properties.* The surface area of the filler can have a significant effect on paper properties. The adsorption of papermaking chemicals such as sizing chemicals, dyes and fluorescent whitening agents on the surface of filler particles or into pores within the particles can cause problems. Figure 11.17 shows the effect of filler surface area on sizing efficiency. In some cases the surface charge of the filler can also be important in that the retention of papermaking chemicals is affected. For example, the use of a cationically-charged filler will increase the retention of anionic papermaking chemicals whereas an anionically-charged filler will

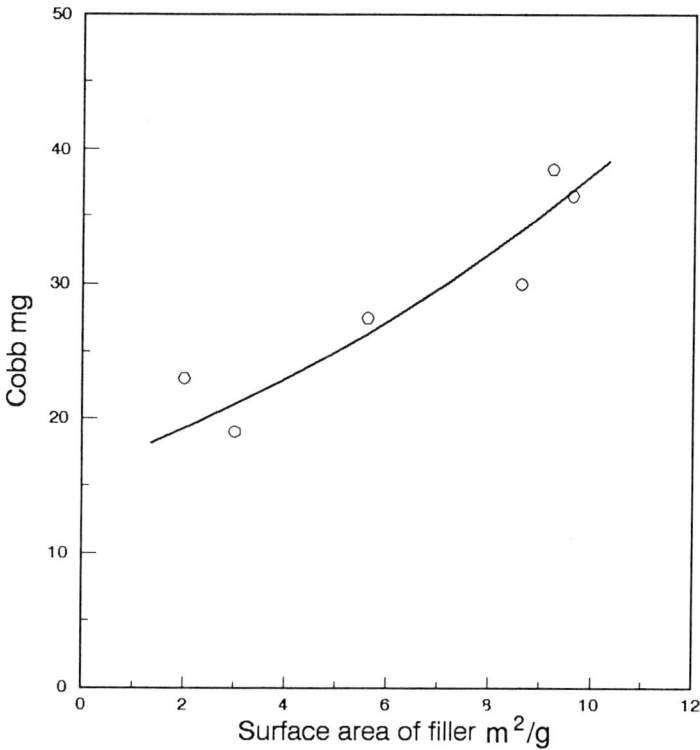

Figure 11.17   Plot of Cobb against filler surface area for calcium carbonate filler at 15% w/w loading. Alkyl ketene dimer size, 50:50 softwood/hardwood Kraft furnish (400 CSF).

possibly reduce retention [54]. There is, therefore, a combined effect: surface area can reduce the effectiveness of papermaking chemicals and surface charge can either increase or decrease their retention. The hydrophobicity of the filler can also be important. For example, talc and other hydrophobic minerals are used to adsorb detrimental pitches to prevent contamination of machine surfaces.

Further to the discussion on the effect of filler–fibre bonding on paper strength, there is some evidence that the strength of attachment of filler particles to the surface of fibres may be influenced by surface properties. In general, coarser particles are preferentially removed as dust (typically in a printing process) from the surface of paper. Removal of particles is, however, reduced by surface treatment of the paper with starch or other polymeric binders. Also, the slightly hydrophobic talc particles appear to be more easily removed than hydrophilic kaolinite or mica particles of similar size and shape [55]. As stated above, these differences in strength of attachment, whilst relevant to paper dusting, are not necessarily relevant to paper strength.

The increasing use of flotation de-inking systems has led to research on the influence of the presence of filler on process efficiency [56]. The surface properties of the filler in terms of adsorption of inks and flotation chemicals are probably relevant but conclusions so far are conflicting.

11.4.2.2 *Light-scattering coefficient.* The light-scattering coefficient of filler as a dry powder is a function of the refractive index, the wavelength of illumination, the physical nature of the particles and their proximity to each other. The light-scattering coefficient increases with increasing refractive index. This is most significant for pigments such as titanium dioxide, which have high refractive indices, but for the majority of fillers the refractive index is close to that of cellulose (1.53). Table 11.10 gives the refractive indices of typical fillers. The light-scattering coefficient also passes through a maximum with increasing particle size. Particle size can refer either to filler particles in

Table 11.10   Refractive indices of typical fillers

Substance	Refractive index
Cellulose	1.53
Titanium dioxide	
anatase	2.55
rutile	2.70
Kaolin	1.57
Calcium carbonate	1.57
Hydrated aluminium oxide	1.57
Silica	1.45
Urea formaldehyde	1.58
Talc	1.57
Alumino-silicate	1.55

air, or to air voids (pores) within filler particles [57, 58]. The optimum size depends on the wavelength of illumination, particle shape and the refractive index [59], but lies in the range 0.4–0.8 μm for fillers such as calcium carbonate and kaolin and is about 0.2 μm for titanium dioxide. Figure 11.18 shows the relationship between particle size and light-scattering coefficient for kaolin as a filler in a beaten softwood sulphite fibre sheet. As the wavelength of illumination increases, the light-scattering coefficient decreases and the optimum particle size increases. Thus for conventional fillers, the light-scattering coefficient at 550 nm is about 10% less than that at 450 nm and the optimum size is about 10% greater. The rate of change of the light-scattering coefficient with wavelength of illumination is a useful indication of the particle size of the scattering units [57].

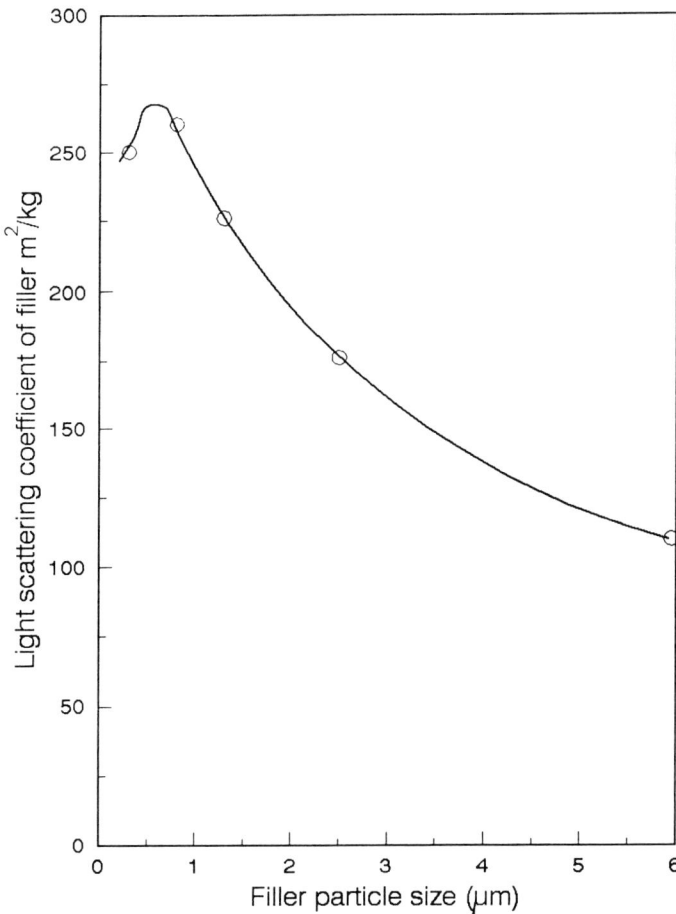

Figure 11.18 Plot of filler light-scattering coefficient against particle size for narrow size distribution kaolin filler at 20% w/w loading in a bleached softwood sulphite (300 CSF) handsheet.

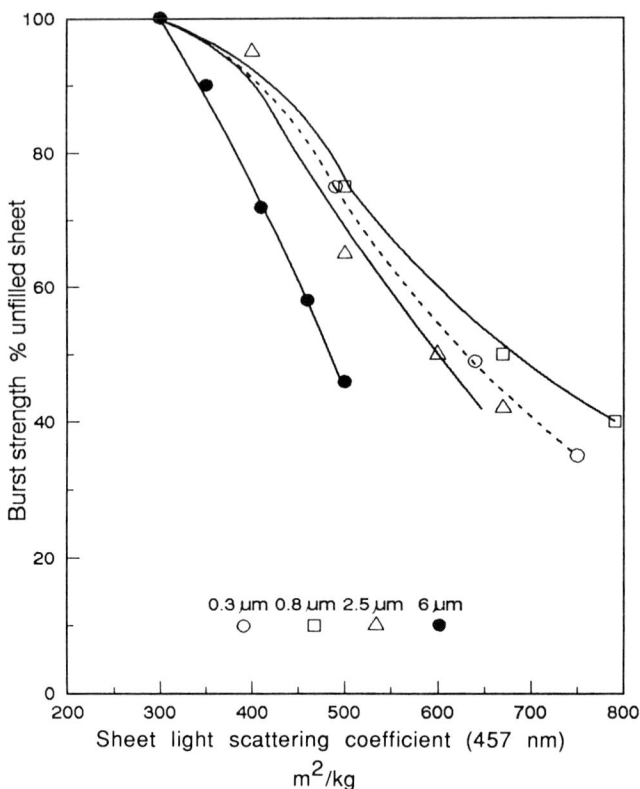

Figure 11.19 Relationship between sheet strength and light-scattering coefficient as a function of filler particle size (kaolin).

The relationship between paper strength and light-scattering coefficient can be important for commercial papers. For non-aggregated fillers, the light-scattering coefficient of a filled sheet at a given paper strength increases as the filler particle size moves closer to the optimum size for light scattering. This is shown in Figure 11.19 using data from the same sheets as used for Figure 11.18.

In paper, the light-scattering coefficient of the filler is not easily separated from that of the fibre [41]. A simple linear relationship of the following form is often assumed:

$$S_{paper} = S_{unfilled\ sheet}(1 - L) + S_{filler}(L)$$

where S_{paper} and $S_{unfilled\ sheet}$ are the light-scattering coefficients of the filled and unfilled paper, respectively, prepared under identical conditions, S_{filler} is the light-scattering coefficient of the filler and L is the weight fraction of filler in the paper. Unfortunately, calculated in this way S_{filler} contains all the

Table 11.11   Comparison of light-scattering coefficients (450 nm) for fillers at about 20 wt% loading*

Filler (see Table 11.3)	Typical uncalendered[†] fine paper furnish $(m^2 kg^{-1})$	Machine calendered[††] mechanical paper $(m^2 kg^{-1})$
Standard kaolin	150	90
Standard chalk	140	120
Fine calcined kaolin	340	290
Coarse alumino-silicate	180	130
Silica	220	150

[†] 50:50 hardwood:softwood bleached Kraft (400 CSF) standard handsheets.
[††] TMP softwood (100 CSF) pilot machine + machine calender.
*Assuming simple linear relationship between the light-scattering coefficients of pulp and filler and the loading.

contributions due to disruption of the fibre network and is therefore dependent on fibre properties and papermaking conditions such as beating conditions, fibre fibrillation, pressing conditions, calendering, filler loading etc.

Table 11.11 gives light-scattering coefficients for a number of fillers in a typical uncalendered fine paper (50:50 hardwood:softwood blend beaten to 400 CSF) and in a machine calendered mechanical paper. The values are significantly lower in the mechanical paper and this indicates the sensitivity of light-scattering coefficient to the physical environment of particles.

Certain fillers such as calcined kaolin and some precipitated alumino-silicates, have relatively high light-scattering coefficients even in a calendered paper and this is not explained by refractive index or by specific gravity differences. The light-scattering coefficient for these fillers contains a substantial contribution from internal pore volume. The pore volume must have a characteristic pore diameter close to the optimum for light-scattering [60]. Some pigments such as fine precipitated silicas, which have high pore volumes, have pore diameters which are characteristically too low to scatter light effectively [61].

11.4.2.3   *Light-absorption coefficient.*

The light-absorption coefficient of a filler is dependent on the bulk chemistry of the filler and, typically, on the presence of coloured impurities, either as particulate matter or as a surface coating, or as elements substituted into a crystal structure. Charge transfer complexes and natural organic materials are a usual source of colour in natural mineral fillers. Synthetic fillers generally have very low light-absorption coefficients. Table 11.12 gives typical light-absorption coefficients at 450 nm for commercial fillers.

When fillers are used in papers that contain fluorescent whitening agents added to the paper furnish, the absorption of UV radiation is very important. Most fillers containing iron or organic contamination in amounts sufficient to

Table 11.12 Typical light-absorption coefficients (450 nm)

Filler (see Table 11.3)	Light-absorption coefficient, $(m^2 kg^{-1})$
Standard kaolin	2.5–3.5
Standard chalk	2.5–3.5
Fine calcined kaolin	1.2–2.0
Coarse alumino-silicate	~0.2
Silica	~0.1
Marble	0.4–0.6

give absorption coefficients of $3–4\,m^2g^{-1}$ at 450 nm, reduce the efficiency of fluorescent whitening agents by about 20% at a filler content of about 20 wt% in the paper when compared with similar fillers with minimal contamination. Titanium dioxide, however, absorbs very strongly in the UV and, at a filler content of 5 wt%, the efficiency of fluorescent whitening agents can be reduced by up to 50%.

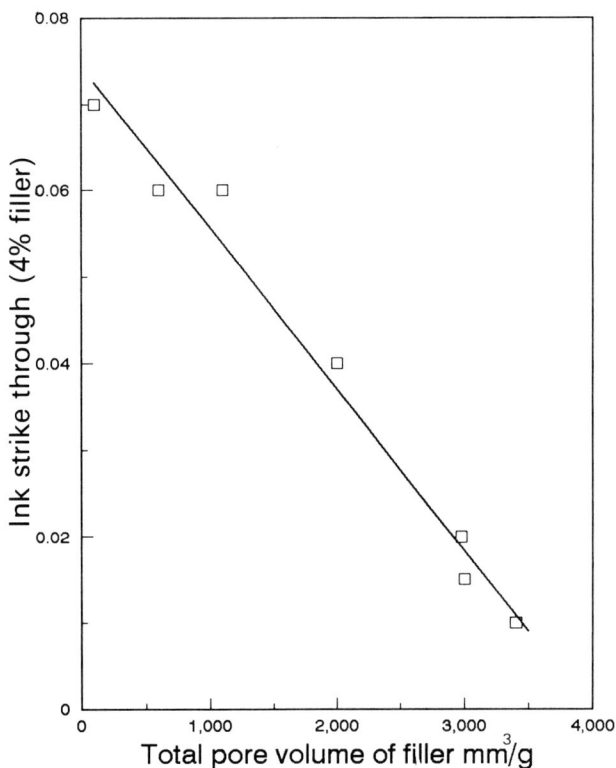

Figure 11.20 Ink strike through for a 40 gsm soft wood TMP sheet containing 4 w/w% filler, plotted against the total pore volume, $mm^3 g^{-1}$ filler. (Measured by Hg porosimetry).

11.4.2.4 *Particle shape and internal porosity.* The effect of filler particle shape and internal porosity with respect to size measurement, retention and filler–fibre interaction have been discussed.

Particle shape is particularly relevant to the air permeability of paper. In general, air permeability decreases with decreasing filler particle size, but also decreases with increasing filler particle aspect ratio [55]. Typically, kaolin fillers with particle aspect ratios of at least 10:1 give up to 50% less air permeability at 20 wt% loading than calcium carbonate fillers with particle aspect ratios of about 2:1. Particle shape is also important in the response of filled papers to calendering, and this applies mainly to the supercalendered papers [55]. Platey particles can be orientated by the forces applied by a supercalender to lie parallel to the paper surface. This greatly improves paper gloss and smoothness. Gloss levels up to 40% (Tappi 75° gloss) can be achieved at 30 wt% loading with kaolin and talc fillers. At the same time, the light-scattering coefficient is decreased and the density of the sheet increases substantially. In fact, the apparent density of the filler in supercalendered paper (assuming that the apparent fibre density is independent of filler loading) is close to the actual filler density.

Internal porosity affects the light-scattering coefficient of the filler, as has been discussed. The porosity also contributes to surface area and this is important in the effect of filler on papermaking chemicals. Porosity, however, also increases the absorption of inks, and this can be beneficial to printing properties. Unlike the light-scattering coefficient, the effect of porosity on ink absorption is not particularly sensitive to pore size [60, 61]. Figure 11.20 shows a plot of ink strike through,[4] against filler pore volume for a range of fillers at 5 wt% loading in a TMP paper at a sheet weight of 40 gsm. Porosity also seems to be important for the modification of paper friction properties by fillers [62]. Highly porous fillers, such as synthetic silicas, significantly increase paper friction presumably by absorbing certain fatty acids and soaps into the internal pore structure.

11.5 General summary

A wide variety of fillers are used in paper, either simply to reduce production costs, or to provide specific paper properties. The mechanisms by which they are retained and by which they influence paper properties are understood to a greater or lesser extent, but there are some general principles which appear to apply to all fillers.

---

[4] Ink strike through $= \log \dfrac{\text{reflectance of unprinted paper}}{\text{reflectance of back of printed paper}}$

Reflectance is measured over a black background. Constant amounts of ink are supplied to give suitable comparison between pigments.

Particle size and shape are obviously very important, and for many pigments the standard measurements of particle size need to be considered in conjunction with a knowledge of shape or degree of permanent aggregation and pore structure, before the size information is used in a discussion of paper properties.

Filler retention is dependent on surface chemistry but, because the papermaking process is unlikely to reach a state of thermodynamic equilibrium, the kinetics, and hence influence of particle size and numbers become very important. Indicative measurements of surface charge, such as zeta potential, provide a guide to potential difficulties in retention rather than to detailed retention behaviour. Comparisons of the retention of different fillers cannot be made on the basis of surface chemistry alone – the effect of actual particle size can be equally important.

In the final sheet, the effect of a filler is dependent not only on intrinsic properties of the filler particles, but also on the disruptive influence of the filler on the fibre network. For example, fine fillers that pack loosely can be very disruptive, reducing the paper strength, and increasing the bulk and the light-scattering coefficient. It is important to appreciate that, where fibre disruption is a major contributor to the influence of a filler, any attempts to overcome deficiencies in important paper properties, notably strength, may have detrimental effects on other paper properties such as the light-scattering coefficient. Some fillers, however, have important intrinsic properties such as a high refractive index, internal porosity or a low specific gravity and these may have effects on paper properties that clearly dominate effects from fibre disruption.

References

1. Magemeyer, R.W., ed. *Pigments for Paper*, Tappi Press (1984).
2. See e.g., Gill, R.A., *Nordic Pulp and Paper Research Journal*, **4** (2) (1989), 120.
3. Allen, T., *Particle Size Measurement*, Chapman and Hall Ltd, 3rd Edition (1981).
4. Jennings, B.R. and Parslow, K., *Proc. R. Soc.*, **A419** (1988), 137.
5. *See* Stratton, R.A. and Suranson, J.W., *Tappi*, **64** (1) (1981), 79; Stark, H. and Eichinger, R., *Wochenblatt f. Pap.*, **23/24** (1984), 871–876; Davidson, R.W., *Tappi*, **57** (12) (1974), 85.
6. Levine, S. and Friesen, W.I., *Flocculation in Biotechnology and Separation Systems*, ed. Y.A. Astia, Elsevier Science Publishers B.V., Amsterdam (1987), 3–20.
7. Hunter, R.J., *Zeta Potential in Colloid Science*, Academic Press (1981).
8. Friend, J.P. and Hunter, R.J., *J. Colloid Interface Sci.*, **37** (1971), 548.
9. Friend, J.P. and Kitchener, J.A., *Chem. Eng. Sci.*, **28** (1973), 1071.
10. Le Bell, J., Bergroth, B., Stenius, P. and Stenlund, B., *Papper och Trö*, **5** (1974), 463.
11. Hall, E.S., *J. Appl. Chem.*, **15** (May 1965), 197.
12. Boluk, M.Y. and Van de Ven, T.G.M., *Colloids and Surfaces*, **46** (1990), 157.
13. Jaycock, M.J., Pearson, J.L., Counter, R. and Husband, F.W., *J. Appl. Chem. Biotechnology*, **25** (1975), 815.
14. Versluys, R.P., *Chemistry of Papermaking Conf. Proc.* (1991), 18.
15. Siffert, B. and Fimbel, P., *Colloids and Surfaces*, **11** (1984), 377.

16. Sanders, M. and Schaefer, J.H., *Tappi Papermakers Conference* (1989), 69.
17. Davidson, R.W., *Tappi*, **57**(12) (1974), 85.
18. Weigl, J., Hlavastch, J. and Scheidt, W., *Wochenblatt f. Pap.*, **13** (1989), 587.
19. Alince, B. and Lepoutre, P., *Tappi*, **66**(1) (1983), 92.
20. Alince, B., *Colloids and Surfaces*, **23** (1987), 199.
21. Fimbel, P. and Siffert, B., *Colloids and Surfaces*, **20** (1986), 1.
22. Jaycock, M.J. and Pearson, J.L., *J. Appl. Chem. and Biotechnology*, **25** (1975), 827.
23. Stark, H. and Eichinger, R., *Wochenblatt fur Pap.*, **23/24** (1984), 871–876.
24. See for example: Theng, B.K.G., *Clays and Clay Minerals*, **30**(1) (1982), 1; Howarth, G.J., Hudson, F.C. and West, J., *J. Appl. Pol. Science*, **21** (1977), 29; *Adsorption from Solution at the Solid/Liquid Interface*, eds Parfitt, G.D. and Rochester, C.H.
25. Krause, T., Schempp, W. and Hess, P., *Deutsche Pap.*, **3** (1984), 127.
26. Hendrikson, E.R. and Neuman, R.D., *Tappi*, **68**(11) (1985), 120; Roberts, K., Kowalewska, J. and Friberg, S., *J. Colloid and Interface Sci.*, **43**(3) (1974), 361.
27. Williams, W.R. and Mather, R.D., *Paper Technology and Industry* (1975), 150.
28. Al-Jabari, M., Van Heiningen, A.R.P. and Van de Ven, T.G.M., *J. Pulp & Paper Sci.*, **20**(9) (1994), J249.
29. Al-Jabari, M., Van Heiningen, A.R.P. and Van de Ven, T.G.M., *J. Pulp & Paper Sci.*, **20**(10) (1994), J289.
30. Middleton, S.R. and Scallan, A.M., *J. Pulp & Paper Sci.*, **17**(4) (1991), J127.
31. Middleton, S.R. and Scallan, A.M., *Nordic Pulp & Paper J.*, **9**(3) (1994), 156.
32. Tam Doo, P.A., Kerekes, R.J. and Pelton, R.H., *J. Pulp and Paper Sci.*, **10**(4) (1984), J80.
33. Britt, K.W., *Tappi*, **56**(10), (1973), 46.
34. Levine, S. and Friesen, W.I., *Flocculation in Biotechnology and Separation Systems*, ed. Y.A. Attia, Elsevier Science Pub., Amsterdam (1987), 3; Lindström, T., *Fundamentals of Papermaking*, eds Baker, C.F. and Punton, V.W., MEP, London (1989), 309.
35. Andersson, K., Sandström, A., Störm, K. and Barla, P., *Nordic Pulp and Paper J.*, **2** (1986), 26.
36. Smoluchowski, M., *Z. Phys. Chem.*, **92** (1971), 129.
37. Van de Ven, T.G.M., *Fundamentals of Papermaking*, eds C.F. Baker and V.W. Punton, MEP, London (1989), 471.
38. Hubbe, M.A., *Tappi J.*, **69**(8) (1986), 116.
39. Hubbe, M.A., *Colloids and Surfaces*, **16** (1985), 249; Pandya, J.D. and Spielman, L.A., *J. Coll. Interface Sci.*, **90**(2) (1982), 517.
40. Gregory, J., *Flocculation in Biotechnology and Separation Systems*, ed. A. Attia, Elsevier Sci. Pub., Amsterdam (1987), 31.
41. Wägberg, L., *Adsorption of Polyelectrolytes and Polymer-Induced Flocculation of Cellulosic Fibres*. Royal Institute of Technology, Department of Paper Technology (1987).
42. Davison, R.N., *Tappi*, **66**(11) (1983), 69.
43. Waech, T.G., *Tappi*, **66** (1983), 137.
44. Bown, R., *Papermaking Raw Materials*, ed. V. Punton, MEP, London (1985), 543.
45. Goodwin, L., *Tappi Papermaker's Conference* (1989), 257.
46. Economu, P., Hardy, J.F. and Menashi, J., *Tappi Papermaker's Conference*, (1975), 117–125; Kenaga, D.L., *Tappi*, **56**(12) (1973), 157.
47. Schott, H., *Tappi*, **54**(5) (1971), 748.
48. Alince, B., Robertson, A.A. and Inoue, M., *J. Colloid and Interface Sci.*, **65**(1) (1978), 98.
49. Alince, B. and Lepoutre, P., *Tappi*, **64**(4) (1981), 135.
50. Bown, R., *Wochenblatt fur Pap.*, **107**(6) (1979), 197.
51. Lindström, T. and Florén, T., *Svensk Papperstidn.*, **87**(12) (1984), R97; Lindström, T. and Florén, T., *Nordic Pulp & Paper J.*, **2**(4) (1987), 142.
52. See Bown, R., *Paper Technology and Industry*, **26**(6) (1985), 280.
53. Alince, B., *Fundamentals of Papermaking*, eds C.F. Baker and V.W. Punton, MEP, London (1989), 495.
54. Bown, R., *Paper Technology*, **31**(7) (1990), 32.
55. Bleakley, I.S., Bown, R. and O'Neill, G.P., *Wochenblatt fur Pap.*, **7** (1987), 298–302.
56. See Schriver, K.E. *et al.*, *Proc. Tappi Pulping Conf.*, **1** (1990), 133; Zabala, J.M. and McCool, M.A., *Tappi J.*, **71**(8) (1988), 62; Letscher, M.K. and Sutman, E.J., *J. Pulp & Paper Sci.*, **18**(8) (1992), 5225.

57. Gate, L.F., *J. Phys. D: Appl. Phys.*, **5** (1972), 837.
58. Alince, B., *Paperi ja Puu – Papper och Tuä*, **8** (1986), 545.
59. Kerker, M., *The Scattering of Light and Other Electromagnetic Radiation*, Academic Press (1969).
60. Bown, R., *Wochenblatt fur Pap.*, **113** (14) (1985), 517.
61. Roeht, W.W., *Tappi*, **49** (6) (1966), 265.
62. Withiam, M.C., *Tappi J.*, **74** (4) (1991), 249.

12 Measurement and control

F. ONABE

12.1 Introduction

Computer-based, on-line process control systems are utilised in virtually every aspect of the pulp and paper manufacturing process today. However, the one major exception to this is the process associated with wet-end chemistry [1, 2].

Recent trends toward faster machine speeds, increased use of twin-wire technology, and the production of paper with a higher ash content have created a need for better first-pass retention [3]. To these trends, the market has responded by developing better retention aids that enhance the papermaker's control over first-pass retention [3].

The primary purpose of an on-line retention monitoring system is to learn more about what is going on at the wet end and what measures are required to maintain stability. Instability at the wet end often characterises itself as a deterioration in product quality, and mistakes in its control cannot usually be corrected later on in the process [2].

Another major purpose of a wet-end monitoring system is to provide real time information during machine trials, for example, the effects of new wet-end additives, or new wires and felts [2]. With real time information, successful trials can be made to produce more useful data in a shorter number of steps and, if unsuccessful, can be stopped sooner with minimal negative effect on production and paper quality [2].

This chapter is intended to provide an overview of the state of the art of on-line measurement and control of wet-end chemistry in the papermaking process.

First, the current status of on-line wet-end chemistry measurements is reviewed. Then the reason why wet-end chemistry process control instrumentation has not advanced as far as in other areas of the paper manufacturing process will be discussed. Recent examples of on-line retention and zeta potential monitoring systems will be introduced and, finally, a perspective for the future methodology for complete automatic retention control systems will be presented.

12.2 Current status of wet-end chemistry measurement

A survey by the Retention and Drainage Sub-committee of the Tappi
Papermaking Additive Committee on the application of instrumentation and
control devices of wet-end chemistry in paper mills was conducted in 1983 [1]
with a view to establishing why on-line measurement and control had not come
into practical use. This survey concluded that papermakers and suppliers
were, at the time, making a wide variety of wet-end chemistry measurements, of
which only three, namely, pH, thick stock consistency and additive flow, were
measured on-line routinely. Freeness, thin stock fibre consistency, zeta
potential and first-pass filler retention sensors were on the market, but had not
been widely used by papermakers at that time.

The survey also indicated that there was a strong need for developing on-
line methods of measuring cationic demand, process stream ash and fines
concentrations, thick stock consistencies, dissolved inorganics, freeness and
drainage, and first-pass retention. The results are summarised in Table 12.1.

Since this survey was conducted several years ago, the response to the
questionnaire will not necessarily reflect the present status of utilisation of on-
line instrumentation. Today, in 1990, a few examples of computerised on-line
retention and zeta potential monitoring devices are available. However, a
complete automatic control system for wet-end chemistry incorporated in the
control loop has not yet been realised.

Table 12.1 Survey of necessary wet-end chemistry measurements [1]

Measurement	Papermakers	Suppliers
Cationic demand, charge, zeta potential, etc.	31	26
Process stream ash contents	9	4
Process stream fines contents	2	1
Thin-stock consistencies	7	0
On-line retention	7	7
Dissolved inorganics (alum, sulphate, acidity, etc.)	12	7
Dissolved organics (Not anionic contaminants)	3	2
Fibre properties (length, flexibility, count)	6	3
Freeness, drainage	8	8
On-line formation	2	0
Furnish air content	0	2
Antioxidant retention	1	2
Surface tension	1	1
Wire/felt deposits	0	1
Rate of ASA hydrolysis/ esterification	0	1

12.3   Problems in measurement and control in wet-end chemistry

There are a number of difficulties associated with on-line measurement and control in wet-end chemistry during the papermaking process. These difficulties can be classified into three factors: colloid chemistry, sensor technology and control technology [4, 5].

12.3.1   *Colloid chemistry factors*

It has been shown [1] that measurements associated with liquids, such as pH and additive flow rates, are relatively easy to carry out compared with those associated with colloidal dispersion and fibre or filler suspensions. In particular, measurements related to surface electric charge, zeta potential or cationic demand can be very difficult. Paper stock running on a high-speed paper machine wire is a system consisting of pulp fibres and fillers as well as a variety of additives undergoing heterocoagulation, and is under dynamic equilibrium with the hydrodynamic shear force or turbulence of the paper machine. Under these conditions, the DLVO theory for a uni-component colloidal system under quasi-static equilibrium cannot be used to explain its colloidal behaviour. Whilst colloidal forces facilitate flocculation, the hydrodynamic shear forces facilitate deflocculation. Paper stock on a high-speed paper machine can therefore be considered to be a dynamic equilibrium between these colloidal hydrodynamic forces. Conventional theories of colloid chemistry are thus not sufficient to explain the behaviour of paper stock on the paper machine.

12.3.2   *Sensor technology factors*

On-line sensors are crucial for maintaining the quality of paper products. Although a wide variety of on-line sensing devices are available for flow and consistency measurement, those for on-line real-time surface charge measurement such as zeta potential or cationic demand await development. Even where measurement with on-line sensors is available, output signals are not necessarily incorporated into the control loop of a control system.

12.3.3   *Control technology factors*

Since the paper stock is a hetero-coagulating system running at high speed, a variety of factors such as surface charge of pulp and fillers, pH and ionic strength of water and additives, and hydrodynamic parameters, affect its colloidal and hydrodynamic behaviour.

Generally, in process control, a mathematical model is required correlating input parameters with output variations of the controlled system. To control wet-end parameters such as retention, drainage and formation by wet-end additives, a mathematical model correlating the additive parameters with the

INPUT   OUTPUT

Ionic Strength ———→   ┌─────────────────┐   Zeta Potential
pH   ———→   │ Mathematical   │   Variation
Cationicity or   │ Model for   │ ———→ Cationic Demand
Anionicity   ———→   │ Surface Charge │   Variation
Molecular Weight ———→ │ Control   │ ———→ Other Variations
                       └─────────────────┘

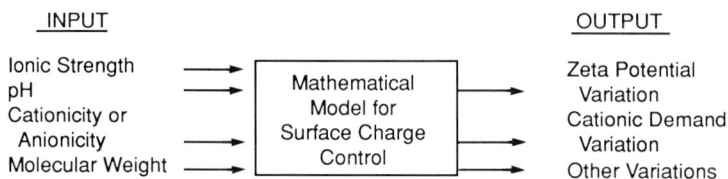

Figure 12.1 Mathematical model correlating the additive parameters with the zeta potential and cationic demand.

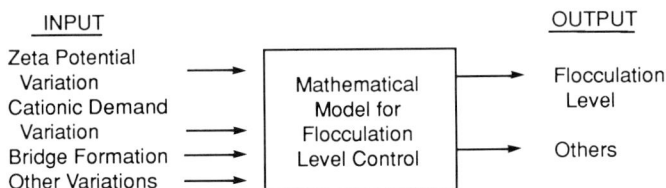

INPUT   OUTPUT

Zeta Potential
Variation   ———→   ┌─────────────────┐   Flocculation
Cationic Demand   │ Mathematical   │ ———→ Level
Variation   ———→   │ Model for   │
Bridge Formation ———→ │ Flocculation   │   Others
Other Variations ———→ │ Level Control │ ———→
                       └─────────────────┘

Figure 12.2 Mathematical model correlating the parameters affecting flocculation with flocculation level.

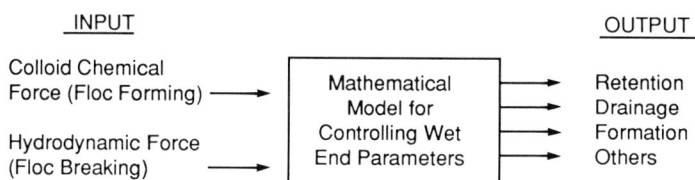

INPUT   OUTPUT

Colloid Chemical   ┌─────────────────┐ ———→ Retention
Force (Floc Forming) → │ Mathematical   │ ———→ Drainage
                       │ Model for   │ ———→ Formation
Hydrodynamic Force   │ Controlling Wet│ ———→ Others
(Floc Breaking)   ——→ │ End Parameters │
                       └─────────────────┘

Figure 12.3 Mathematical model correlating the colloid chemical and hydrodynamic forces with the wet-end parameters.

zeta potential and cationic demand is first required (Figure 12.1). Secondly, a mathematical model correlating the parameters affecting flocculation such as zeta potential, cationic demand and the bridge forming capability of polymeric additives is needed (Figure 12.2). Finally a mathematical model correlating colloid chemical and hydrodynamic force with the wet-end parameters is required (Figure 12.3). Since, normally, these input–output relationships are unknown, it is not an easy task to construct either a mathematical control model or a simulation model.

At the present time, the control of stability of the wet end has been accomplished only by accumulating empirical data from the paper machine and using it to predict the variance and external fluctuations.

12.4 On-line wet-end chemistry instrumentation

It is well known in papermaking practice that some of the important parameters which need to be monitored and controlled on-line are pH, aluminium, retention and surface charge of stock component. The most recent example of instrumentation for the monitoring of surface charge [6] is believed to create new possibilities for total control of wet-end chemistry.

The following are the on-line wet-end chemistry sensors available [1].

12.4.1 Dissolved inorganics

The paper industry makes extensive use of on-line pH sensors. Conductivity and specific ion electrodes are also available. The latter are used for selectively measuring specific cations and anions of interest in wet-end chemistry. Detection of specific anionic species is important because of their contribution to the cationic demand of the stock. In the practical use of these electrodes it may also be necessary to control temperature, pH and ionic strength.

12.4.2 Dissolved organics

Dissolved anionic organic materials known as 'anionic trash' are believed to be the principal contaminants which interfere with cationic wet-end additives by neutralising their cationicity. The only method currently available for obtaining a quick estimate of the total dissolved organics in a process stream is to employ an on-line total organic carbon analyser (TOC). It has been successfully used to monitor the TOC in the flow of paper machine effluent to a waste water treatment plant [1]. For dissolved total anionic substances, the cationic demand measurement is also required.

12.4.3 Flow rates

Magnetic flow meters have been mainly used to monitor additive stream and process flow rates and they provide mill operators with accurate and precise flow information. Recently, flow meters based on ultrasonic or electromagnetic principles have become available.

12.4.4 Solid consistency measurements

On-line measurement and control of pulp and fibre consistencies by optical means is a particularly important tool for retention monitoring. A conventional consistency device does not measure the weight of fibres or fillers in suspension directly, but measures the shear force exerted by the flowing stock suspension or changes in its optical properties. The selection of the kind of device depends on the consistency of the suspensions. Above 1% consistency, the most commonly employed sensors utilise the principle that the shear force exerted by the flowing stock suspension is directly related to its consistency [1]. The shear force in this case is independent of flow, pressure and turbulence variations.

Optical consistency sensors are commonly used as non-contact sensing devices and are reported to be adequate, particularly for thick-stock consistencies. Successful applications for pulp slurries as high as 7% consistency have been reported [1]. It is anticipated that optical sensors will increase in importance in line with the trend to high consistency papermaking technology.

Below 1% consistency, the force exerted by the flowing stock is too low for shear sensors to work effectively, and optical devices utilising the principles of absorption, transmission, scattering or depolarisation of light by suspended solid particles are used. Laser-optical consistency devices have recently been developed for on-line retention monitoring systems. These will be described later.

There are also devices which use the principle that consistency and the absorption of ultrasonic energy are correlated. This relationship is very sensitive to changes in the degree of beating of the stock and the amount of air in the slurry.

Most equipment measures consistency indirectly by non-contact means and does not provide a fundamental measurement of the consistency of the paper stock. The following are the consistency devices currently available.

12.4.4.1   *EUR-Control LOWCON Sensor.* This device is based on the principle that fibres display optical activity and that the degree to which they rotate polarised light is proportional to their consistency. The light attenuation characteristics of the fillers and other non-polarising additives are allowed for by a built-in compensation circuit.

12.4.4.2   *STFI Fibre Length Analyser.* This device is an on-line fibre length analyser and operates on the principle that the particles passing through the cell attenuate the light beam and cause its intensity to fluctuate. The detector signals contain information on both the particle size distribution of the fibres and also of the fibre concentration. It is possible to discriminate the fibres from the smaller sized fines and fillers.

12.4.4.3   *Kajaani FS-100 Fibre Length Analyser.* This device is an off-line fibre length analyser and operates on the principle of flowing a dilute pulp suspension through a narrow capillary. The fibres are thus separated, and are identified and their length measured by polarised light.

12.4.5   *Other measurement techniques*

12.4.5.1   *Freeness measurement.* On-line freeness testers are available and their principal use is in refiner control. A modified version of the freeness tester is also claimed to be suitable for on-line drainage monitoring.

12.4.5.2 *Surface charge measurement.* There are a variety of off-line zeta potential measuring devices on the market. Some of the currently available sensors are also claimed to be suitable for making on-line electrokinetic measurements [17]. Although they can be used as monitoring devices, they are not sufficiently adaptable to be used as real-time control devices.

12.4.5.3 *System 3000 automated electrokinetic analyser.* This device is capable of measuring specific conductance, pH and electrophoretic mobility simultaneously. It also provides information about the mobility distribution function, turbidity, settling velocity and particle size. The overall system consists of an on-line sampler, a sensor unit and a computer. Since the system utilises the microelectrophoresis principle, the data are not available on a real-time basis. A sequential control circuit is thus built in to be automated by microcomputer.

12.4.5.4 *Zeta reader.* This device also utilises the microelectrophoresis principle. On-line sampling is conducted in flowing suspensions at regular intervals and electrophoretic measurements are made. Electrophoretic behaviour can be visualised on a CRT display.

12.4.5.5 *MBS-8000 Sensor.* This device utilises a novel measuring principle, that is the electrokinetic sonic amplitude measurement technique (ESA). It is capable of rapid measurement of complete furnish suspensions and has future potential for on-line electrokinetic measurements.

The most recent advanced surface charge monitoring system will be described later.

In addition to the above devices, it has also been proposed that cationic demand can be used as an indication of the surface charge [7, 8]. However, there are theoretical questions which must be answered if this method is to be applied to paper stock [9].

12.4.4.6 *First-pass retention measurements.* There are a variety of devices capable of measuring the first pass retention on a laboratory scale. Some of these have recently been claimed to be capable of on-line, first-pass retention measurement. The M/K Systems Retention Meter provides real-time, continuous information about a paper machine's first-pass retention performance using light transmission data from the headbox and white water. It is reported that first-pass retention values obtained using this device show a good correlation with gravimetrically determined values for a wide range of furnishes and consistencies [1].

12.5 Process control in wet-end chemistry

The process of paper manufacture has inherent variability which must be minimised if the machine operation is to yield a uniform, high quality paper product with efficient utilisation of pulp, filler, wet-end additives, manpower and energy. The objective of process control in wet-end chemistry is to maintain process variables such as retention, drainage and formation at certain set points or within well-defined limits of variation [10]. For instance, the zeta potential has an acceptable control range or critical value to optimise wet-end parameters (Figure 12.4) [11].

When a control is under manual control, the operator has to monitor the trend of the data points. It is then necessary to anticipate a break through the limit lines so that proper action can be taken in advance to keep the process values within limits. This is called 'feed-forward control'. A hypothetical feed-forward loop for cationic demand control is shown in Figure 12.5 [5].

However, depending on the relative process stability, the accuracy of data

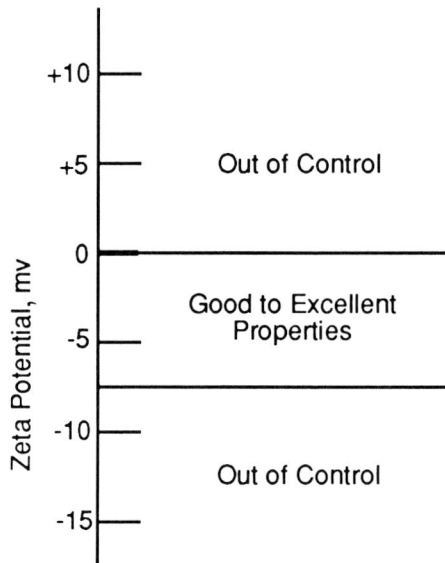

Figure 12.4   Optimum control range of zeta potential [11].

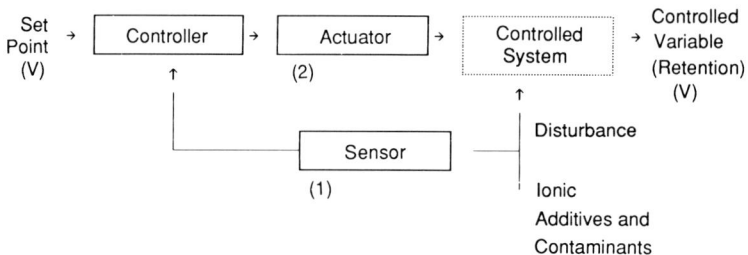

Figure 12.5   A hypothetical feed-forward loop for cationic demand control in wet-end system: (1) cationic demand measurement, (2) retention and flow rate control.

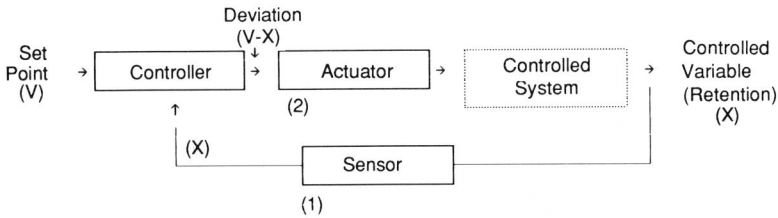

Figure 12.6 A hypothetical feedback loop for retention control in wet-end system: (1) first-pass retention measurement; (2) retention and flow control.

points and the narrowness of the limit lines, the operator may only be able to take corrective action after the variable has strayed outside the process limit. This is called 'feedback control'. A hypothetical feedback loop for retention control is shown in Figure 12.6 [5].

Process parameters in wet-end chemistry are still not under computer control, but are under manual control. In automatic process control systems there is a control loop consisting of three basic components: sensor, controller and actuator. The sensor transmits signals to the controller, which in turn compares measurements with predetermined set values. Next, the controller transmits signals to the actuator so that the controlled system will maintain its stability. The interdependence among these components is shown schematically in Figure 12.7. A hypothetical complete automatic retention control system might have a cationic demand control loop and a retention control loop [1].

The sensor is at the heart of any control loop, but, as yet, sensors capable of measuring directly the important parameters such as retention or surface charge are not easily available. For complete and automatic control of wet-end chemistry, a mathematical model describing the fluctuation of input parameters (e.g. pH, ionic strength etc.) and variations of output parameters (e.g. zeta potential, retention etc.) is required [12].

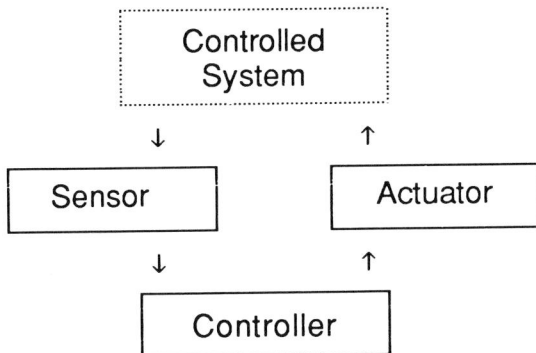

Figure 12.7 Basic components of control loops.

12.6   On-line retention monitoring systems

One of the main aims in automatic wet-end chemistry control is the control of retention. A variety of on-line continuous retention monitoring systems are now available from manufacturers in Scandinavia. It is claimed that, by adopting these tools, higher speeds, fewer breaks, better runnability, a higher production rate and uniform paper quality can be achieved. The operating principles for these systems are based on the measurement of consistencies of dilute suspension in the headbox and white water system by laser-optical means. Although they are computer-controlled on-line retention monitoring systems, they are not incorporated into the automatic control loop. Therefore, complete automatic retention control has not been realised at present. Characteristics of representative systems are described below.

12.6.1   *Kajaani RM-200 system (Finland) [3, 13]*

Kajaani already has the LC-100 system on the market. This is an optical transmitter capable of measuring suspensions in the low consistency range of 0–1.5% [14]. A characteristic of the RM-200 system is that it is capable of measuring the filler and fibre consistencies separately using their different polarisability values. The sensor for this purpose is illustrated in Figure 12.8. Laser technology and integrated internal calibration mechanisms make this continuous measurement accurate and stable.

The system can measure simultaneously both filler and fibre consistencies at three locations in the wet-end system. An example of an installation of this system on a twin-wire paper machine is shown in Figure 12.9. The sensors

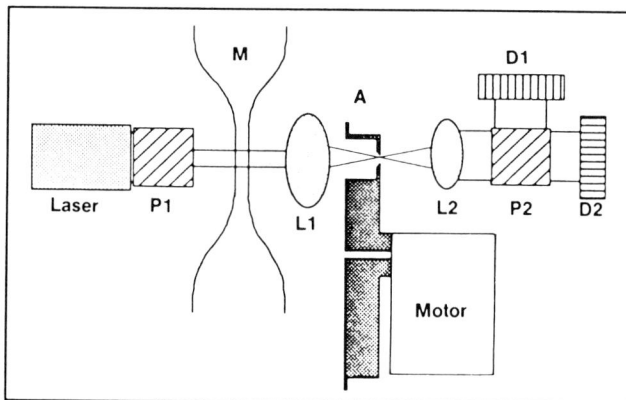

Figure 12.8   Sensor for measurements of total solids and filler consistency [3]. P1: polariser; M: measurement cell; L1: Fourier lens; A: rotating aperture disc; L2: collimating lens; P2: polariser; D1 and D2: photodiodes.

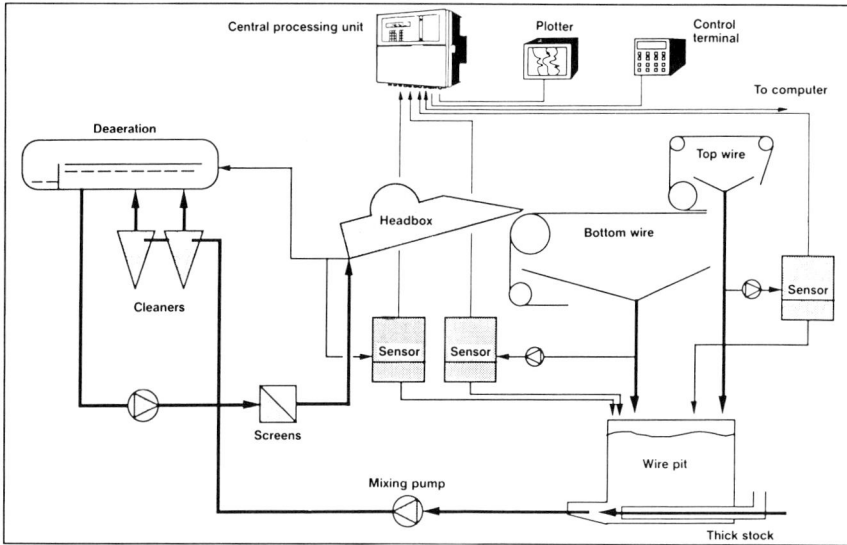

Figure 12.9   Kajaani's retention monitoring system [3].

transmit the consistency values to the central unit which then calculates retention continuously.

When the sensors are installed in the headbox and the wires, the filler and fibre consistencies are measured *in situ*. The consistency data are used to calculate the wire retention, i.e. the fraction of headbox stock retained in the web after water removal. The continuous wire retention measurement gives information which allows dosage control of retention aids and fillers.

12.6.2   *Chemtronics 4000 system (Sweden) [2, 15, 16]*

This system also determines retention by measuring the consistencies of fibre and filler in the headbox and in the white water system. The standard system consists of a main unit and an operator station (Figure 12.10). The main unit, consisting of a sampling system, a sensor package and a computer is installed near the wet end of the paper machine (Figure 12.11). The operator station is usually installed in the control room. A computer controls the measurement cycle which takes normally four minutes and calculates total solids, filler and retention.

The sensor package consists of a laser wave sensor and a microwave sensor (Figure 12.12). The laser sensor measures the fluid properties such as turbidity, scattering and depolarisation of polarised light. Depolarised and polarised light signals are measured for transmitted and scattered light. This gives four kinds of signals from one sample. Since these four signals differ according to

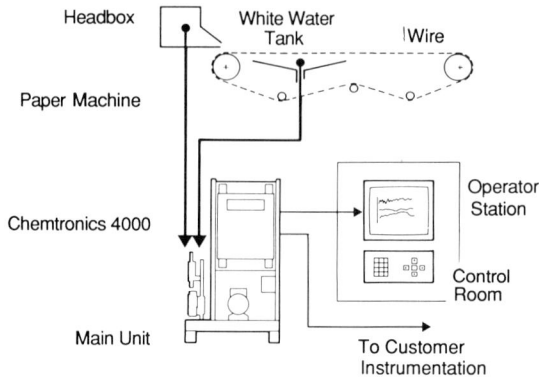

Figure 12.10 Chemtronics retention monitoring system [16].

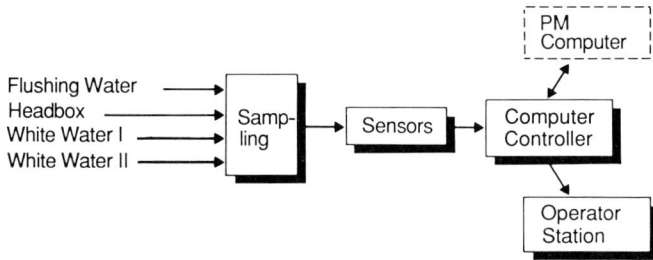

Figure 12.11 Sequence from sampling to monitoring [16].

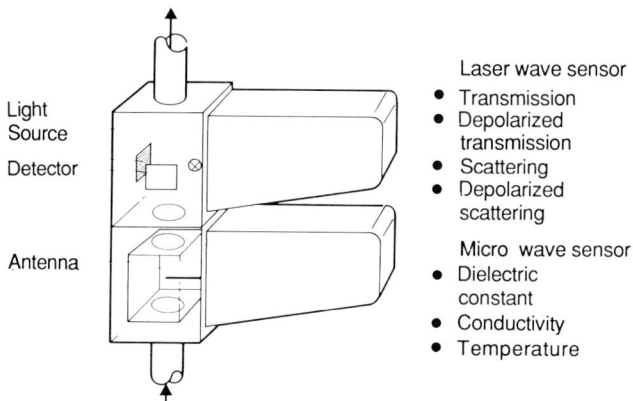

Figure 12.12 Sensor package consisting of laser wave sensor and microwave sensor [16].

the kind of fibre and filler, it is possible to obtain filler consistency and total solid consistency separately using one sensor. The microwave sensor measures the properties of the suspending medium such as dielectric constant, conductivity, and temperature.

The above properties can be used independently or in combination depending on the complexity of the furnish, for accurate filler and fibre determinations. The computer controls the sampling cycle and calculates the consistencies and retention from the raw sensor signals.

Although this system works as an independent retention monitoring system, it can be incorporated into a closed retention control loop by adding an adaptive control algorithm to its software.

As for communication of monitored data, the internal communication in the mill is based on serial data transfer over a four-wire cable which can be extended to the main unit or be connected to satellite sensors or other measurement and control systems. External communication is also possible over a data modem, which is a useful tool for troubleshooting or for recalibration of the system.

12.7   Monitoring of on-line surface charge

The idea of on-line zeta potential measurement has been proposed so far on the basis of the streaming potential method (Figure 12.13) [17] and bias detection of charged flow direction in the electric or magnetic field (Figure 12.14) [17]. Recently an on-line device utilising the streaming potential principle has been announced [7].

The 'ZETA DATA' [7, 8] is an on-line zeta potential device developed by Penniman in 1989 in cooperation with Bayer AG. This device utilises the streaming potential principle and also simultaneously measures the conductivity and temperature of the furnish. These values are used to calculate the zeta potential using the Helmholtz–Smoluchowski equation. The overall

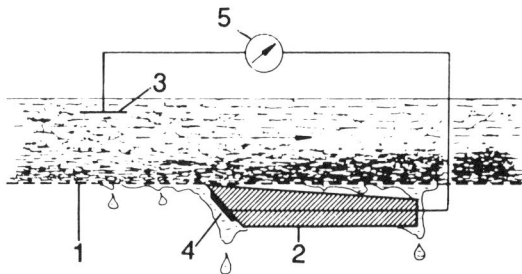

Figure 12.13 Electrode arrangement for continuous measurement of streaming potential on paper machine. 1:wire; 2:foil; 3:upper electrode; 4:lower electrode; 5:measuring device [17].

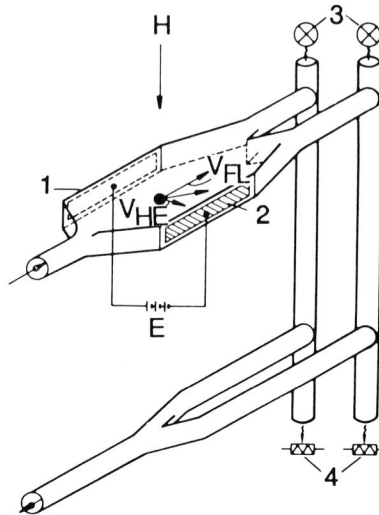

Figure 12.14 Continuous electrophoretic and magnetophoretic zeta potential measurement utilising bias detection of charge flow direction. 1 and 2: electrodes; 3: bulbs; 4: light sensitive elements [17].

operation consists of the automatic formation of a pad, through which white water is passed. The streaming potential is then measured across the pad. This device can be used over a consistency range of 0.2–0.6% and over a freeness range of 0–800 CSF [7, 8]. Using this system, the rate of addition of wet-end additives can be adjusted so as to maintain the zeta potential at an optimum level. The system has the possibility of being used as a means of attaining complete automatic retention control.

12.8   Methodology for complete automatic retention control

To attain complete automatic wet-end chemistry control incorporating control loops using the monitoring systems and control methodology described previously, a strategy has to be adopted. Scott [1] proposed the approaches described in the following section for fines retention control.

12.8.1   *Design of control loops for retention control*

The following three generalisations, available from experimental findings have to be accepted in order to design control loops:

(a)   First-pass retention is critical to the wet-end chemistry performance of most additives and components.

(b) Retention aids are necessary to achieve acceptable levels of first-pass fines retention on paper machines of medium to high speed.

(c) Dissolved and colloidal anionic substances in a wet-end system tend to interfere with the effectiveness of cationic additives.

From these three generalisations, a model for a first-pass retention control strategy has been presented and is shown in Figure 12.15. Although this model has not yet been made the basis of a control system, it may come into practical use in the near future instead of the conventional manual control of retention.

The strategy can also be described as a combination of control loops (Figure 12.16). In the first loop (B1–B2), the anionic contaminants coming into the system are measured and neutralised in the thick stock part of the stock preparation system. This is a loop for cationic demand measurement and control using the feed-forward control principle. In the second loop (A1–A2), the first-pass retention value is measured and the retention aid addition rate is adjusted if a change in retention is required. Information from the stock

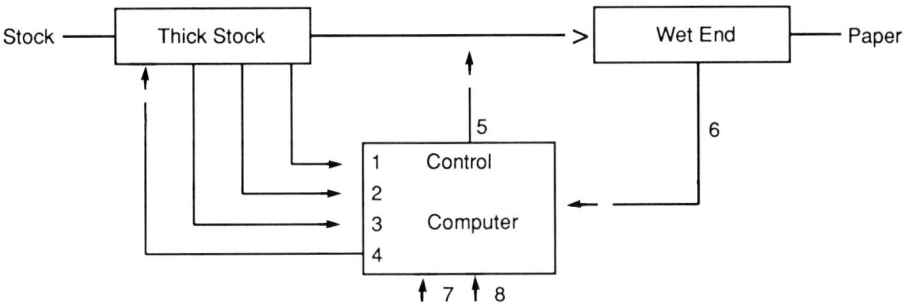

Figure 12.15 Model for computer-controlled first-pass retention control strategy. 1: fines mass flow rate sensing; 2: filler flow rate sensing; 3: cationic demand sensing; 4: charge neutraliser addition; 5: retention aid addition; 6: first-pass fines retention sensing [1].

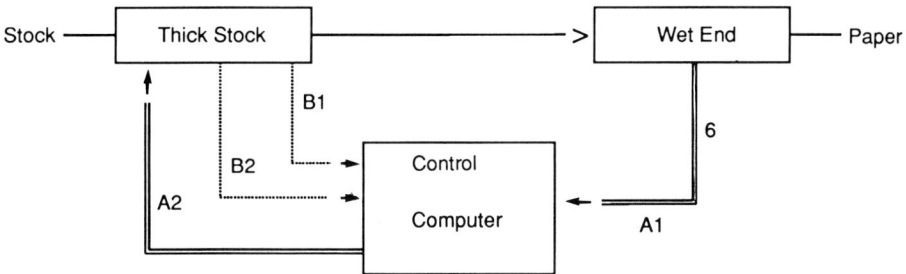

Figure 12.16 Control loops for the model in Figure 11.15. A1–A2: first-pass retention control loops; B1–B2: cationic demand control loops.

preparation system is also fed to the first-pass retention control loop. This is a control loop based on the feedback control principle. The third invisible control loop of this model is the feedback of paper property information to the process computer. The data based on this information will be fed into the retention control loop.

The characteristics of this model are the requirements of a continuous first-pass retention measurement and the active use of a retention aid to control first-pass retention on a real-time basis.

12.8.2 *Selection of monitoring systems and sensors*

Since the actual papermaking system is a complex quasi-colloidal system undergoing hetero-coagulation, accurate measurements of the parameters are not easy. A minimum of seven types of on-line measurements are required:

1. Cationic demand level
2. Charge neutraliser addition rate
3. Thick stock flow rate
4. Thick stock fines retention
5. Filler addition rate
6. Retention aid addition rate
7. First-pass fines retention

These measurements can be conducted on-line using existing devices on the market. The flow rates of charge neutraliser, filler and retention aids can be measured on-line using existing flow meters. Retention data can be measured on-line using the recent advanced retention monitoring systems of Kajaani or Chemtronics. The cationic demand level can be measured by the automated colloid titration technique or streaming current detector [18]. On-line zeta potential measurement is also possible by the ZETA DATA which measures electrokinetic and conductivity properties of the bulk paper stock utilising streaming potential principles.

When considering the hardware aspects of wet-end control systems, it is theoretically possible to construct an automatic retention control system using existing on-line sensing devices.

12.8.3 *Development of computerised retention control systems*

To switch from a conventional manual retention control system using existing on-line retention monitoring methods to a fully computerised automatic retention control system, it is necessary to combine a loop for electric charge control and a loop for retention control. The software environment in the form of a control algorithm and mathematical model for the wet-end system has then to be developed.

The problem that remains to be solved is the design of mathematical models describing these complex wet-end systems. Since the construction of complete and precise mathematical models for the wet-end system is almost impossible, the input–output relationships for the control model have to be obtained from the accumulated data in the mill. That is, it is advisable to collect manually, or on-line, on a frequent basis, the values for headbox and tray ash and consistency, stuffbox ash, cationic demand, additive addition rates, etc. Then it is necessary to find out the correlation between the first-pass retention values and these accumulated values. Through these trial-and-error methods, one can set up and improve the quality of mathematical models for the wet-end system.

12.9   Applicability of new sensing devices and techniques

Although a variety of new sensing devices and techniques are available, those applicable to a complex system like paper stock, consisting of colloidal particles and large solid particles under hydrodynamic shear, are very few. However, some of them have potential applications to the wet-end system in the future.

12.9.1   *Flow visualisation techniques*

The recent advance of image analysis techniques in combination with small-sized CCD cameras has made flow visualisation possible using a pattern recognition principle. For the analysis of flocculation phenomena under dynamic conditions, techniques using strobo-photography and image analysis have been proposed [19]. If a correlation between floc image characteristics and retention characteristics could be established, the image calculation data obtained from the flow visualisation technique have potential applications for retention control. A variety of problems such as consistency limit, opacity, light refraction and the construction of mathematical models have still to be solved.

12.9.2   *Opto-electric sensing devices*

These are combinations of laser light and optical fibres capable of being applied to a wide variety of measurements such as image, pressure, flow rate, and temperature. Since laser light is coherent and monochromatic, the detailed behaviour of pulp and filler in suspension may be measured with extreme precision.

12.9.3   *Ultrasonic sensing devices*

Since ultrasonic waves travel in any kind of medium with different velocities depending upon the type of medium, they have a variety of applications in measurement and control. Ultrasonic-based level meters, flow meters, and consistency meters are widely used today in industrial process control. A zeta potential device using the ultrasonic principle is also on the market.

12.9.4 *Magnetic sensing devices*

The measurement of zeta potential using the electro-magnetic principle has been proposed [17]. The development of a device for cationic demand measurement is also being investigated.

12.9.5 *Ion-electrode sensing devices*

There are a variety of ion-electrode sensing devices on the market, which utilise electrochemical principles. The selective measurement of ionic characteristics of wet-end additives in paper stock is a possible development.

12.10   Concluding remarks

On-line automatic control is an ultimate goal in wet-end chemistry. Above all, the focal point is the control of fibre and filler retention. Current on-line retention and surface charge monitoring systems can be used as trouble shooting tools, as a method for making machine adjustment and for controlling and optimising wet-end stability.

Although on-line continuous retention monitoring systems, have now come into practical use, the automatic retention control system incorporated into feed-forward or feedback control loops has not yet been realised. To attain this goal, a mathematical model describing colloid and hydrodynamic interactions associated with wet-end chemistry of papermaking systems has to be formulated. Furthermore, the development of sensors capable of measuring the surface charge of fibres and fillers in paper stock continuously on a real-time basis is vital. For this purpose, a variety of innovative charge measuring systems have been proposed in recent years as seen in the special section dealing with the wet-end charge analysis in Tappi Papermaker's Conferences in 1993.

In addition, a newly emerging electronic-based information technology system 'multimedia' is expected to provide a drastic change in overall control technology. Multimedia is an integrated digital information technology system consisting of multicolour, movie and sounds. These were made possible by high speed CPU and high capacity memory devices of personal computers. This tremendous development of personal computers will lead in general to down-sizing from minicomputers to personal computers in control systems as well as increased flexibility of control.

References

1. Scott, W.E., *Tappi J.*, **67** (11) (1984), 72.
2. King, C.A., Smith, W.B. and Stein, W. *Advanced Topics in Wet End Chemistry Seminar*, 103, Tappi (1987).
3. Kortelainen, K., Norkelainen, J., Huttunen, J. and Lehmikangas, K., *Tappi J.*, **72** (8) (1989), 113.
4. Onabe, F., in *Sheishi-Kagaku* (Paper Science), ed. T. Kadoya *et al.* (in Japanese), Chugai-Sangyo-Chosakai (1982).
5. Onabe, F., *Japanese J. of Paper Technol.*, 1986 (in Japanese).
6. Dillen, S., *World Pulp and Paper Technology*, Sterling Pub. Int. Ltd (1990), 249.
7. The ZETA DATA catalogue (Paper Chemistry Laboratory, 1989).
8. Penniman, J.G., *Revue A.T.I.P.*, **43** (10) (December 1989), 513.
9. Onabe, F., *J. Japan Wood Res. Soc.*, **28** (7) (1982), 445.
10. Smook, G.A., *Handbook for Pulp & Paper Technologists*, Tappi, CPPA Joint Textbook Committee (1982) Ch. 243.328.
11. Penniman, J.G., *Paper Trade J.*, (Jan. 1978).
12. Onabe, F., *Japan Tappi.*, **39** (3) (1985), 20.
13. Kajaani RM-200 catalogue (Kajaani Automation, 1989).
14. Connolly, K.P., *Tappi J.*, **70** (3) (1987), 89.
15. Stein, W., *Tappi J.*, **70** (4) (1987), 63.
16. Chemtronics 4000 catalogue (Eka Nobel, 1989).
17. Sack, H.W., *6th Fundamental Research Symposium*, Oxford (1977).
18. Onabe, F. and Sauret, G., *Revue A.T.I.P.*, **36** (5) (1982), 213.
19. Onabe, F., *9th Fundamental Symposium*, Cambridge (1989).

13 Practical applications of paper chemistry

C.O. AU

13.1 Introduction

This chapter covers certain aspects of the practical applications of paper chemistry. It is not intended as an extensive review.

Retention is one of the most important processes of the wet-end operation. Apart from the cellulose fibres, many functional additives are added to the stock prior to drainage. In order to provide the desired properties, they have to be efficiently retained. An optimum level of retention can reduce the cost of chemicals, increase machine output and decrease the load on the saveall system. The practical application of retention aids is therefore included in the discussion.

Among the many functional chemicals used at the wet end, size and starch are two of the most common. Their main purposes are to control liquid penetration and increase dry strength respectively. As a result of a more in-depth understanding of the nature of their interaction with cellulose fibres, new types of sizes and starches have been developed which have made the use of soap size and native starch at the wet end almost obsolete. The application and problems associated with these new chemicals will be considered.

As pitch depositions have always been a problem to papermakers, and they are frequently related to chemicals found at the wet end, it is informative to review how both fundamental and practical studies have helped to control, if not solve, this problem.

Although extensive research has been carried out in wet-end chemistry, there are still many areas in which the present knowledge has failed to explain and solve the phenomena observed and the problems encountered. Such limitations and difficulties, together with their causes, will be discussed at the end of this chapter.

13.2 Retention

Studies concerning the mechanism of retention aids have been carried out by many researchers and an extensive review has been presented recently [1]. The subject is also discussed in detail in Chapters 3 and 5. It is now understood that, at least as far as one-component systems are concerned, they can function mainly via bridging, charge neutralisation or by the patch charge mechanism. The exact mechanism of operation is dependent upon molecular weight and charge density. Bridging is favoured by polymers of high molecular weight and low charge density, while the patch charge model is favoured by high charge density but is relatively independent of molecular weight. Their behaviour towards turbulence has also been studied [2], and the classification of 'soft' and 'hard' flocs has been proposed. These understandings are vital in considerations such as the addition point of polyacrylamide and polyethylenimine retention aids. Polyacrylamides are understood to function via the bridging mechanism and are usually introduced late in the system, such as after the centri-screen. This is because these 'hard' flocs are only resistant to a short period of turbulence. Extensive exposure to conditions of high shear will tend to break down these flocs. Furthermore, the extent of reflocculation is found to be incomplete and this will lead to poor retention. Polyethylenimine, on the other hand, induces the formation of 'soft' flocs via the patch charge model. These flocs, although easily redispersed when exposed to high shear, are believed to reflocculate completely when the shear is removed. As a result, the choice of addition point of this type of retention aid is not as critical.

One of the major limitations when using chemicals which can function as retention aids is their influence on sheet formation. Their ability to behave as polyelectrolytes and induce flocculation can, in some cases, deteriorate the formation of the sheet. This is more critical with the manufacture of fine paper since its look-through appearance can be crucial. As a result, papermakers very often cannot afford to run their machines at high levels of retention. Fortunately, there are many other mechanical means of optimising formation and they have been reviewed recently [3].

From a chemical supplier's point of view, this limitation presents opportunities for the introduction of a new generation of retention aids. As a result, two-component systems have been developed. Generally they consist of a high molecular weight cationic polyelectrolyte which induces the initial flocculation, followed by very small anionic inorganic particles to consolidate these flocs. Examples of such systems include the combinations of cationic starch with an anionic silica sol and cationic polyacrylamide with modified bentonite. The exact mechanisms are not entirely clear but acceptable suggestions have been put forward [4,5]. In addition, fundamental flocculation studies [6,7] show that they do give rise to smaller flocs at comparable degrees of flocculation. These 'tighter' flocs are believed to contribute to the maintenance of good formation at high levels of retention.

Studies concerning the adsorption behaviour of polymers onto fibres in the presence of simple electrolytes have been carried out and are reported in the literature [8,9]. This has an important practical application because of the presence of various types of interfering ions at the wet end. It is now understood that, for sufficiently high molecular weight polymers, adsorption onto fibres increases with an initial increase in the electrolyte concentration. A further increase in concentration will reduce the level of polymer adsorption, due to a decrease in electrostatic interaction between fibres and polymer. On the other hand, if the polymer is of medium molecular weight, there is a large increase in adsorption with increasing electrolyte concentration. This is explained by the decrease in polymer size at high electrolyte concentration, making the interior structure of the fibres more accessible. These findings underline the importance of the characterisation of polymers for retention aid purposes in different electrolyte environments. Besides considering the basic requirements of the mechanisms of flocculation, it is now also possible to consider the environment in which the polymers will operate. High molecular weight polymers seem to be favourable in systems where the level of conductivity is high. This will ensure that the adsorbed polymers are located on the outer surface and are therefore more likely to be available for bridging. On the other hand, when the conductivity is low, comparatively lower molecular weight polymers may be sufficient for the required level of flocculation. This will therefore reduce the risk of over-flocculation and hence the risk of deterioration in formation.

One of the classic examples of a solution concerning a retention problem is described here. A neutral fine paper mill, using a two-component retention system consisting of cationic starch and anionic silica sol, was found to suffer from periods of poor retention [10]. This had caused high usage of retention chemicals. Extensive investigation work was carried out to try to locate the cause of the problem. It was noticed that the level of calcium ions in the backwater was at an unusually high level, and, as high levels of calcium ions are known to decrease the adsorption of cationic polyelectrolyte, it was anticipated that the adsorption of cationic starch had been reduced. This belief was confirmed by the high level of COD found in the backwater. Since cationic starch was functioning as part of the retention system, these calcium ions must have been responsible for the poor level of retention. The next step was to locate the source of the calcium ions. It was noticed that the concentration of dissolved calcium ions in a sealed bottle of backwater increased with time. This suggested that some sort of bacterial activity was taking place, probably by acid-generating bacteria, which caused the dissolution of calcium carbonate. After carrying out the appropriate routine work to pinpoint the growth of bacteria in the system, it was decided to introduce additional biocide dosage points at these locations. The result was that the level of calcium ions gradually returned to normal and retention remained at its usual level with no extra usage of retention chemicals. This example demonstrates that fundamental knowledge of the adsorption of polyelectrolyte (cationic starch) and other basic understandings have been applied to solve practical problems.

13.3 Starch

Starch is widely used in the paper industry for a number of applications, including wet-end addition, surface treatment and coating. The discussion which follows is mainly concerned with starch as a wet-end additive.

Starch, when used as a wet-end additive, has been variously described as functioning as a 'beater size', a 'beater adhesive' or a 'chemical hydrater'. Despite this discrepancy in terminology, its function as a dry-strength additive is now incontrovertible. One of the drawbacks of using starch as a wet-end additive is its low level of retention, which is believed to be due to its low affinity for cellulose fibres [11,12]. This has led to the introduction of cationic starches, the idea for which originated from that of aminocellulose derivatives which were used for the purpose of improving the dyeing characteristics of cellulose fibres. Nowadays, the use of starch at the wet end is almost exclusively confined to these cationic derivatives. Its high level of retention has been proved to be due to its inherently greater affinity towards the negatively charged cellulose fibres [12,13]. Comparative studies into the retention of non-ionic and cationic starches by papermaking fibres using carbon-14 labelling techniques have been carried out [14]. Figure 13.1 shows the differences in levels of starch retention and the effects of pH and beating.

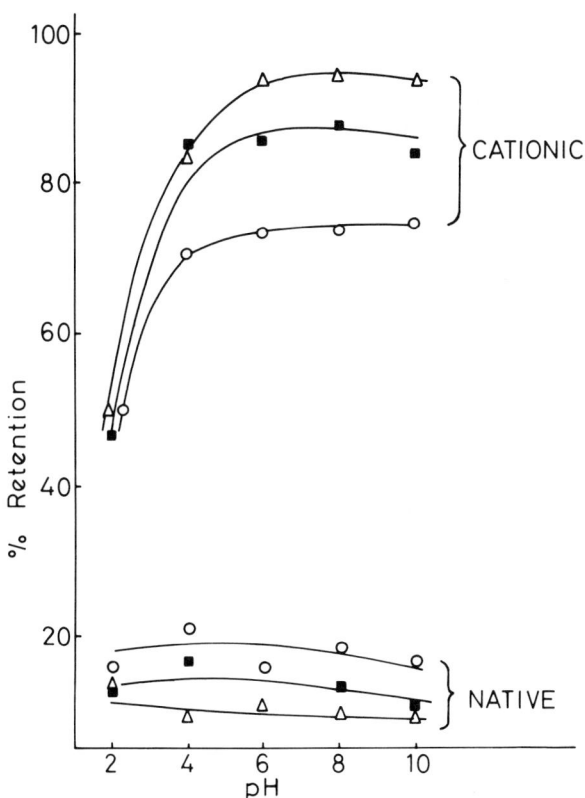

Figure 13.1 The effect of beating and pH on the retention of native and cationic starch by a bleached softwood sulphate pulp. ($\triangle$ 0 minutes; $\blacksquare$ 30 minutes; $\bigcirc$ 60 minutes.)

Explanations for these results have been suggested [14] and will not be discussed here. Nevertheless, they confirm the general belief that much higher retention can be achieved with cationic starch, as compared with non-ionic starch.

From the practical point of view, there is a general belief that the point of addition of cationic starch has an important effect on its function. Addition of the cationic starch early in the system, e.g. at the machine chest, normally improves the dry sheet strength to a larger extent than adding it later, e.g. before the centri-screen. The latter mode of addition, however, can improve the overall retention of the system. The exact cause of this is not entirely certain but many plausible explanations have been suggested. The version given here is based on a difference in retention of cationic starch. It has been suggested that the adsorption of cationic starch onto anionic substrate is both rapid and irreversible [12,15]. Studies regarding the distribution of starch on pulp fractions indicate that the adsorptivity ratio between fines materials and fibres is in the region of 5:1 [16]. This is also discussed in Chapter 6. It follows that when cationic starch is added early in the system, where the fines content is relatively low, adsorption onto fibres is believed to be predominant. As fibre retention is high, more cationic starch is retained by them so as to contribute to bonding and therefore to a stronger sheet. On the other hand, when cationic starch is added late in the system, where the fines content is high, adsorption onto fines materials is believed to be predominant. As retention of fines materials is lower than that of fibres, less starch is retained and thus strength improvement is reduced. The higher overall retention as a result of late addition of cationic starch is probably due to its ability to behave as a polyelectrolyte which, to some extent, helps retention via the bridging mechanism. In a recent publication [17] a novel method of adding strength aid to classified pulp components was reported. Adsorption of strength aids onto long fibres was found to give the strongest sheet, and this was explained by a shift in the mechanism of tensile failure. However, the possibility of different levels of retention of strength aids by different fractions during drainage was not ruled out.

One of the most important aspects regarding the practical application of starch, which cannot be over-emphasised but unfortunately is very often over-looked, is the need of a reliable and efficient starch cooker. Many of the major problems concerning the use of starch are related to problems with the cooker, for instance the feeding of starch and the extent of cooking. This has become much more important since the introduction of cationic starch, because it is known to be able to affect other system parameters such as retention. The traditional method of batch cooking and the use of intermediate storage tanks has, in numerous cases, proved to be insufficient and unreliable. Firstly, the extent of cooking of the starch can be variable due to the batch-wise nature of this process. Incompletely cooked starch can produce different effects [18] and overcooking can, of course, lead to a certain degree of degradation. Secondly, the storage of cooked starch is always susceptible to bacterial attack, leading to degradation. Furthermore, the length of the period of storage can vary depending on the runnability on the machine. The importance of this aspect can be demonstrated in the following example. A fine paper mill using a two-component system consisting of cationic starch and anionic silica sol as retention aid was experiencing problems with unstable retention. The cyclic nature of this variation was noticed to be of a constant frequency, which coincided with the frequency of their batch cooking of cationic starch. The problem was eventually solved by cleaning their intermediate starch storage tank at more regular intervals. This undoubtedly reduced the level of bacterial activity and stable retention level has since been restored.

Problems like these have led to the introduction of continuous steam-jet cookers, which are able to complete the cooking of starch at high temperature and pressure in a short period of time. The cooked starch is then pumped directly to the machine after dilution to a predetermined concentration. Both of the potential difficulties mentioned above can therefore be overcome. Naturally, such continuous cookers are not exactly trouble-free. They do rely heavily on a number of services supplied by the mill, for instance stable steam, water and air pressure. However, reliable cookers have become available nowadays as a result of the experience gathered through extensive operation in mill conditions followed by modification and development work. An example of such a state-of-the-art design of a continuous steam-jet cooker is shown in Figure 13.2.

Figure 13.2   Continuous steam-jet cooker. (Photograph courtesy of Roquette (UK) Ltd.)

13.4 Sizing

The sizing of paper with both rosin/alum and synthetic sizes such as alkyl ketene dimer and alkenyl succinic anhydride has been studied by many researchers in the past two decades. Current understanding regarding their mechanisms has been covered in Chapters 8 and 9 and will not be discussed here. Instead aspects of some practical application and difficulties will be considered.

The rosin size commonly used today is in the form of an anionic dispersion. The use of traditional soap size is now restricted only to a small number of mills because:

(a) acid dispersion exhibits a better tolerance of high pH levels (up to pH 6);
(b) lower Cobb values can be achieved with the acid dispersion;
(c) sheet strength deterioration is less due to a lower alum demand.

These different behaviours have been explained by the difference in the reaction mechanisms of soap size·and dispersion size with alum [19].

In general, alum is added in the mixing chest or machine chest, followed by the introduction of size in the machine chest or level box. With this mode of addition, sometimes known as reverse sizing, a better control of pH can be achieved. It has also been reported to provide an increase in sizing efficiency and a reduction in deposit problems [20].

One of the disadvantages of anionic dispersions is the need for cationic assistance for retention. This can be explained by its inherently anionic nature and thus a lack of electrostatic interaction with cellulose fibres. Logically this has led to the development of different types of cationic rosin dispersions. Their cationic character is believed to be provided by, for example, alum or polyaluminium chloride. Consequently some sort of 'self-retaining' property is incorporated. To demonstrate this property, an example is described here. During a machine trial of one such cationic dispersion in a waste mill making testliner etc., poor sizing was initially observed. This was explained by the high level of anionic disturbing substances in the short circulation system. In order to overcome this problem, an attempt was made to increase the cationic dispersion to an excess level so as to 'mop up' these substances in the backwater. After a short time, a sharp decrease in backwater solids was noticed. The dosage was then returned to an optimum level and good sizing efficiency was obtained.

Alkyl ketene dimer (AKD) and alkenyl succinic anhydride (ASA) are now the most commonly used synthetic sizes. One of the most important practical aspects regarding their application is retention. This is particularly true for ASA because it is known to be extremely susceptible to hydrolysis. Apart from reduced sizing efficiency, hydrolysed sizing material is also a major cause of problems with deposits. It has been reported that the use of a small amount of alum, incorporated into the ASA emulsion, can reduce both size usage and deposits problems. Although the exact mechanism of this effect is not known, an acceptable explanation has been given [21]. Nevertheless, it is understood that the incorporation of alum in ASA emulsions has become common practice.

As the importance of size retention is recognised, cationic size promoters have been developed to improve sizing efficiency, particularly in the case of AKD. These materials can be added separately or incorporated into the size emulsion. Their exact mode of action is not yet fully understood, but their effects are in most cases satisfactory. However, an excess level of these resins is known to be detrimental to the action of the retention system. Furthermore, fundamental studies have shown that the amount of size (AKD) in the final sheet necessary for sizing is extremely low compared with the levels generally added or retained in industrial practice [22].

The methods of application of the two sizes (ASA and AKD) are found to be different. ASA, being more suspectible to hydrolysis, is generally added quite late in the system, e.g. the fan pump. This mode of addition can avoid hydrolysis and give good size retention. AKD, on the other hand, can be added to the thick stock if the mixing at this point is good. Both ASA and AKD should be introduced, ideally, before the addition of filler. This is believed to maximise the interaction between size and cellulose fibres, while minimising the consumption of size by filler particles.

The benefit of understanding the heat curing requirement for AKD-sized paper can be demonstrated in the following example. When poor sizing is encountered, it is useful to compare the cured and uncured Cobb values. If they are both higher than specification, the problem probably originates at the wet end, e.g. poor size retention. On the other hand, if only the uncured Cobb is high while the cured Cobb is within specification, it is likely that the problem lies in the drying section, e.g. an air-lock in the dryer cylinders. One classic example of this happened in a waste mill manufacturing hard-sized container middles. Their sizing problem was found to be only associated with uncured Cobb. This was explained by an insufficiency in heat input for the necessary curing action to take place. As a result, a steam box was installed with the aim of increasing the web temperature for final cure. The problem then disappeared.

13.5 Deposit control

Deposition in papermaking is one of the areas in which an understanding of paper chemistry has contributed to the control of this potentially serious problem. If it is allowed to continue, it can precipitate loss of production and product quality. Nowadays, factors such as the increase in machine speed and closure of backwater system, and the use of recycled fibres and high yield pulps such as CTMP have accentuated the need for more stringent control of the situation.

The types of deposits generally encountered in a papermaking system can be roughly divided into organic, inorganic and biological. However, in most cases, they will turn out to be complex combinations of more than one of these basic types. As biological deposition originates from a different background and is itself a vastly complicated subject, it will not be discussed here. In fact, only wood pitch deposition will be considered, as the understanding of it is more comprehensive, than, for example, white pitch and size-related depositions.

A large number of papers has been published on the subject of wood pitch and its deposition. Fundamental studies [23] have led to the understanding that wood pitch consists of free fatty acids, resin acids, fatty acid esters of glycerol and sterols, and unsaponifiable materials such as sitosterol and betulinol. The relative amounts of these components were found to be dependent upon different wood species, with the notable exception that only softwoods contain significant amounts of resin acids.

The solubility property of wood pitch in organic solvents has given rise to a very useful method for extraction [24] and, if followed by suitable chromatographic techniques [25], the characterisation of its individual components can be achieved. Currently, the most widely used solvent for this purpose is dichloromethane. A list of pitch contents of some of the pulps as determined by dichloromethane extraction [26] is given in Table 13.1.

It can be seen that the percentage of dichloromethane extractables varies significantly with different pulping processes, wood species and degrees of bleaching [27]. Furthermore, the variations within each individual pulp type are believed to originate from differences in ageing or seasoning of wood before pulping [28], bleaching process and the efficiency of the washing process. In summary, information resulting from these studies has contributed

Table 13.1 The dichloromethane extractable contents of various wood pulps

Pulp types	Percentage dichloromethane extractables	
	Range	Typical
CTMP bleached	0.1–0.4	0.3
CTMP unbleached	0.3–0.8	0.5
TMP bleached	0.3–0.6	0.5
TMP unbleached	0.5–1.2	0.8
SGW bleached	0.3–0.8	0.4
SGW unbleached	0.8–1.7	1.0
SW sulphate bleached	0.02–0.1	0.05
SW sulphate unbleached	0.10–0.20	0.1
HW sulphate bleached	0.10–0.50	0.2
HW sulphate unbleached	0.4–1.5	0.5
SW and HW sulphite bleached	0.1–1.5	0.3

the first step towards controlling pitch content in pulps, and therefore has made it possible to minimise its deposition potential.

Studies have also been carried out to investigate the correlation between wood pitch in pulps and the problems with deposition at the paper machine. It has been revealed [29] that the concentrations of colloidal pitch droplets, as measured by a haemocytometer counting technique, were higher with troublesome pulps than those with non-troublesome ones. It was also suggested that pitch does not cause problems until the free pitch within a system has agglomerated [30]. An extensive study of the various mechanisms of pitch deposition has been published [31]. The mechanisms described include resin transfer from fibres, fines and fire deposition, resin transfer to felts at press nips, and deposition of colloidal resin with hydrodynamic shear, evaporation, creaming and coalescence. Based on these and many other findings, an in-depth understanding of the process of pitch deposition has been developed, and, together with it, various products for controlling this problem.

The chemicals used for treatment of deposits range from papermakers' alum to sophisticated water soluble polyquaternary amines and cationic polymer coated kaolin clay. However, the basic approach is one of three types, namely dispersion, adsorption and flocculation. The dispersion concept is based on the belief that free pitch does not cause a problem until it has agglomerated [30]. So a dispersant is added to stabilise the pitch particles by protective colloid action and these are eventually washed from the system. The adsorption concept, on the other hand, depends on the inherent interaction between pitch and inorganic minerals. Talc, the most common mineral used for this purpose, is added for adsorption of colloidal pitch. These talc/pitch entities will then be retained and carried through the system as part of the product. Finally, the flocculation concept functions via a similar mechanism as retention aids in the sense that pitch particles are flocculated by what is generally known as a pitch fixing agent, and retained in the sheet during drainage. In general, very high molecular weight polymers are found to be most efficient for this purpose.

However, all these approaches possess their own disadvantages. The dispersion method will obviously suffer from the closure of the backwater system because the pitch does not become incorporated into the product. Concentrations of dispersed resin can increase to levels at which agglomeration may take place. Also, when dispersants are initially applied to a system with a severe deposits problem they can cause detachment of existing deposits and generate problems downstream. For the adsorption approach, the filler content of the final sheet will increase and therefore strength properties can deteriorate. Also the removal of these potentially troublesome talc/pitch entities depends upon the level of retention. Poor retention can cause an accumulation in the short circulation loop. Lastly, with the flocculation approach over-flocculation can occur and sheet formation will deteriorate in a manner similar to retention aids.

Apart from wood pitch deposition, there are other types of non-biological deposit, such as white pitch deposit. Unlike wood pitch, which has been extensively studied and well-documented in the literature, understanding and information concerning white pitch is scarce. It is understood that white pitch is generally found in paper mills which are recycling coated broke in which the coating binder is some form of latex polyvinylacetate or styrene–butadiene binder. From experience, a change in parameters such as an increase in binder level, a higher ratio of calcium carbonate to clay in the coating pigment, elevation of pulping temperature and system pH, variations in the level of retention and an increased use of coated broke will increase the propensity for white pitch deposition. Unfortunately, there are no genuinely specific chemicals or methods to control white pitch and the general approaches are along the same line as wood pitch. It is therefore felt that more studies are needed in this area to provide a more thorough understanding of the problem. In fact, this is also true for deposit problems related to synthetic sizes. With the use of AKD and ASA rapidly increasing, there is an urgent need for a more basic understanding of this new type of deposit problem. A solution to this problem is likely to accelerate the rate of conversion from acid to neutral papermaking.

13.6 Difficulties and limitations of practical applications

The successful conduct of the progress of wet-end chemistry is possible only if there are frank and unrestricted discussions and cooperation between researchers, papermakers and suppliers. Working together as a team they can improve the efficiency of this important stage of the papermaking process. They will also be equipped with a comprehensive understanding of the operation and be able to solve problems which arise, in a logical manner.

Undoubtedly, there is now a very comprehensive understanding of many aspects of wet-end chemistry, but the practical application has lagged some way behind. This is partly because the freedom to investigate new methods is very often limited by the requirement that machine production must not be affected.

Suppliers are ideally placed to fill this gap between researchers and papermakers. They ought to have a comprehensive understanding of their products and the chemistry behind them. Furthermore, having to deal with papermakers on a day-to-day basis, they are familar with the practical applications.

Conflicts of interests have in some cases slowed down the progress of wet-end chemistry. However, it must be said that these 'partnerships' have also generated many fruitful results which have made the understanding and application of wet-end chemistry what it is today.

Practical application of fundamental understandings is sometimes difficult to achieve due to the inadequacy of good laboratory simulation tools. Because of the nature of the papermaking operation, it is often very difficult to test out new chemicals and techniques extensively on the paper machine. Therefore comprehensive studies often have to rely on work carried out in the laboratory, simulating machine conditions. As an example, one of the most widely used piece of laboratory equipment in this area is the Dynamic Drainage Jar. It has the advantage of being able to measure retention under a predetermined level of turbulence. However, it does have two main disadvantages in that mat formation and backwater recirculation features are not available, as they are on the paper machine.

However, despite its name, the Dynamic Drainage Jar was not designed for measuring drainage. Whether it is suitable to use this equipment for drainage testing has been discussed [32, 33] and the issue is still unclear. This leaves a huge gap in the study of drainage in the laboratory because conventional equipment such as the Schopper Reigler tester and the Canadian Standard Freeness tester are not realistic as they do not provide any level of turbulence. Various modified versions of the drainage tester have been developed [34, 35], but their use is not widespread enough for their reliability to be fully verified.

A pilot paper machine seems to be the answer to fulfil the requirement of an ultimate simulation tool. Unfortunately modern production machines have been designed to run at high speeds and this has left the slow speed of most pilot machines exceedingly unrealistic for studies such as retention. For pilot machines which are able to run at high speeds, their drying capacities are always limited and this has produced doubts as to their suitability in the study of, for example, dewatering aids and the curing properties of sizes.

Probably due to these difficulties and limitations, the advancement of theories in wet-end chemistry has lagged behind the development of new chemical products and techniques. Many of them are already being widely used and they have proved to be superior to those which they replaced. Unfortunately, understanding of their mechanisms of action is still very vague and contradicting explanations have been put forward. These include, for instance, the role of promoter resins in AKD sizing, and the use of alum in neutral systems for retention, drainage and sizing enhancement. It is hoped that an improved working relationship between researchers, papermakers and suppliers and the design of good simulation tools will change this situation in the future.

References

1. Hubbe, M.A., 'How do retention aids work?', *Tappi Proceedings, Papermaker's Conference*, (1988), 389–398.
2. Unbehend, J.E., 'Mechanisms of "soft" and "hard" floc formation in dynamic retention measurement', *Tappi*, **59** (10) (1976), 74–77.

3. Kufferath, W., 'Different wet end devices and formation on a Fourdrinier', *Paper Technology*, June (1989), VI/23-VI/32.
4. Moberg, K., 'Microparticles in wet end chemistry', *Tappi Notes, Retention and Drainage Short Course* (1988), 65–86.
5. Lowry, P.M., 'Wet end balance through the use of multi-component retention systems', *Tappi Proceedings, Papermaker's Conference* (1988), 231–234.
6. Wågberg, L. and Lindström, T., 'Some fundamental aspects on dual component retention aid systems', *Nordic Pulp and Paper Research Journal*, no. 2, (1987), 49–55.
7. Onabe, F. and Sakurai, K., 'Application of flow visualisation technique in wet end chemistry', *Proceedings of 9th International Fundamental Research Symposium*, Cambridge, England, B.P.B.I.F., London (1989), 219–249.
8. Lindström, T. and Wågberg, L., 'Effects of pH and electrolyte concentration on the adsorption of cationic polyacrylamide on cellulose', *Tappi*, **66** (6) (1983), 83–85.
9. Wågberg, L. and Ödberg, L., 'Polymer adsorption on cellulosic fibres', *Nordic Pulp and Paper Research Journal*, **4** (2) (1989), 135–140.
10. Street, G., *Proc. PIRA Conference, Chemistry of Neutral Papermaking*, Slough (1987).
11. Waters, J.R., 'A technique for the evaluation of starch as a wet end additive', *Tappi*, **44** (7) (1961), 185A.
12. Marton, J. and Marton, T., 'Wet end starch: adsorption of starch on cellulose fibres', *Tappi*, **59** (12) (1976), 121–124.
13. Harvey, R.D. *Retention of cationics starches*, *Tappi*, **68** (3) (1985), 76–80.
14. Au, C.O., The Study of Retention and Function of C-14 Labelled Cationic and Non-Ionic Starch in Paper, Ph.D. thesis, U.M.I.S.T., England (1987).
15. Wågberg, L., Ödberg, L., Lindström, T. and Aksberg, Y., 'Kinetics of adsorption of ion-exchange reactions during adsorption of cationic polyelectrolytes onto cellulose fibre,' *J. Coll. Interf. Sci.*, **123** (1) (1988), 287–295
16. Marton, J., 'The role of surface chemistry in fines-cationic starch interaction', *Tappi*, **63** (4), (1980), 87–91.
17. Stratton, R.A., 'Dependence of sheet properties on the location of adsorbed polymer', *Nordic Pulp and Paper Research Journal*, **4** (2) (1989), 104–112.
18. Casey, J.P., 'Beta sizing with starch', *Paper Ind.*, **26** (10) (1945), 1277–1280.
19. Marton, J., 'Fundamental aspects of the rosin sizing process: mechanistic difference between acid and soap sizing', *Nordic Pulp and Paper Research Journal*, **4** (2) (1989), 77–80.
20. Beatty, J.E., 'Optimising acid and alkaline sizing system', *Tappi Proceedings, Base Stock Seminar*, Atlanta, (1985), 38–40.
21. Strazdins, E., 'Theoretical and practical aspects of alum use in papermaking', *Nordic Pulp and Paper Research Journal*, **4** (2) (1989), 128–134.
22. Roberts, J.C., 'The mechanism of alkyl ketene dimer sizing of paper', *Tappi*, **68** (4) (1985), 118–121.
23. Sjostrom, E., *Wood Chemistry: Fundamentals and Applications*, Academic Press, New York (1981).
24. Tappi Test Method, T204 om-88.
25. Douek, M. and Allen, L.H. 'Kraft mill pitch problems', *Tappi*, **61** (7) (1978), 47–51.
26. Jour, P., to be published, Department of Bleaching Chemicals, Eka Nobel, Surte, Gothenburg, Sweden (1989).
27. Farley, C.E., 'Causes of pitch problems and a laboratory method of evaluating control agents', *Tappi Proceedings, Papermaking Conference*, Chicago (1977), 23–32.
28. Allen, L.H., 'Characterisation of colloidal wood resin in newsprint pulps', *Colloid and Polymer Sci.*, **257** (1979), 533–538.
29. Allen, L.H., 'Pitch particle concentration: an important parameter in pitch problems', *Transactions of the Technical Section*, **3** (2) (June 1977), 32–40.
30. Parmentier, C.J., 'Electronic microscopic observation of pitch problems associated with closed water systems', *Tappi*, **56** (10) (1973), 80–83.
31. Allen, L.H., 'Mechanisms and control of pitch deposition in newsprint mills', *Tappi*, **63** (2) (1980), 81–87.
32. Penniman, J.G. and Olson, C.R. 'Using the Britt Jar to measure drainage', *Paper Trade Journal*, 15 April (1979), 34–36.

33. Penniman, J.G., 'More on the Britt Jar drainage test', *Paper Trade Journal*, 15 September (1979), 41–42.
34. Abson, D., Bailey, R.M., Lenderman, J.A., Nelson, J.A. and Simons, P.B., 'Predicting the performance of shear-sensitive additives', *Tappi*, **63** (6) (1980), 55–58.
35. Davison, R.W., 'A new vacuum pulsating drainage procedure for determining fines particle retention', *Tappi*, **73** (8) (1989), 121–127.

Index